The Spiritual Anatomy

The Spiritual Anatomy

The Church Body and Its Relativity to the Human Body

Our heavenly Creator has left behind His indelible Handprints that point to His divine Plan and Purpose for our lives as the Body of Christ. Although we are many members; we are one in Christ. Our ultimate shared work as believers is to glorify God through the redemptive virtue of the Holy Spirit.

Jolly Glover

© 2011©®

Written for the spiritual and methodically matured Christian believer.

Copyright © 2011 ©® by Jolly Glover.

Library of Congress Control Number: 2009911241
ISBN: Hardcover   978-1-4415-9277-4
              Softcover   978-1-4415-9276-7

All rights reserved. This book is protected by the copyright law of the United States of America. The scanning, uploading, and distribution of this book via the Internet or via any other means without permission of the publisher is illegal and punishable by law. Please purchase only authorized electronic editions, and do not participate in or encourage electronic piracy of copyrighted materials. Your support to the author's rights is appreciated. This book may not be copied or reprinted for commercial gain or profit. The use of short quotations or occasional page copying for personal group study is permitted by written request. Permissions will be granted upon request.

Scripture quotations taken from the Holy Bible, American Standard Bible © 1983, 1987 by Thomas Nelson, Inc. Used by permission.

Scripture quotations reflect the author's added emphasis.

Please note that the writing style of this book capitalizes certain pronouns in scripture that refers to the Father, Son, and Holy Spirit and differ from some publishing styles.

Medical terminology and definitions are referenced and used by permission from www.MedicalDictionary.com

This book was printed in the United States of America.

To order additional copies of this book, contact:
Xlibris Corporation
1-888-795-4274
www.Xlibris.com
Orders@Xlibris.com
66747

Contents

Acknowledgments ix
Introduction 11
Anatomy of the Spiritual Body 13
God Effect.. 15
Kingdom Dynamics 17
Mind-set and Purpose 19
Wealth and Riches within the Kingdom 22
What Kind of Harvest Do You Expect? 25
Kingdom Order.. 27
Purpose of the Spiritual Anatomy....................................... 30
What Does God Look Like?.. 32
Names and Titles of God 34
Spiritual Body.. 37
The Denominational Divide Dilemma 39
The Essence of Life... 41
Human Comparison 43-45
The Mind Is an Amazing Creation 49
Being Spiritually Balanced 51
Another Form of Being Spiritually Bipolar 55
Questions that cause separation 57
Should There Be Division within the Body?
  "The Denominational Divide Question"......................... 59
Face of Lies, Deceits, and Darkness 61
Our God Is a God of Love and Order 63
Spiritual and Natural 65-70
How Does This Anatomy Relate to Us as
  Corporate Believers in Christ? 72
Spiritual and Natural 74-77
Within the Anatomy of the Spiritual Body Is a Soul......... 79
Spiritual and Natural 81-83
What takes place within the Spiritual Womb?................... 97
Each Member Is Important.. 108
Foreign Elements that attack the body............................. 110

Transformation Factor 117
Be Good to the Body 120
Becoming Spiritually Healthy 122
The Great Physician 123
Conventional medicine or Alternative medicine? 125
Scientific Theories 128
Kingdom Come Mentality 131
Our Armor of God 133
Coming Together of the Body 134
Unity within the Body 136
The Magnificent Work of the Body of Christ 138
Fit within the Pattern after the Order of Melchizedeck 140
It's Better to Please God 142-144
God Is Love 145-148
God Made Sex to Be Good 149-158
God's Judgment! How Should We Perceive It? 159-161
How Should We Express Love? 162-158
This Is How the Church Body Should Work as One 169-176
How Far Shall We Believe What the Pastor Says? 177-178
How Should We React to Spiritual Demonstrations? 179-181
Righteous Conduct In the Church Body of Christ 182-188
Was Nicodemus Saved? 189-193
Marvel not! You must be born again! 194
What Must I Do to Be Saved? 195-197
These Shall Be the Signs of Those Who Believe 198
Some Traditions Are Worth Keeping 200
Logic Model to Kingdom Building 201
Kingdom Building 203
Kingdom Building: Wrap-Around Model 205
My prayer for the beloved people of God. 206
Index 207

Dedication

This is a love offering to God who is the light of my life . . .
To my family and friends who shared their love and support, as I agonized to give birth to my ministry.

In loving memory of
my deceased natural father, Fred A. Glover, and to my father in the gospel, the late Bishop Robert W. McMurray.

Acknowledgments

This book has been a labor of love. I have felt driven to complete this assignment to assist the Body of Christ in discovering a new awareness; as to the possibilities of corporate fellowship, and empowerment through spiritual collaborative involvement. I pray that the readers of this book will gain a better perspective in the area of "Kingdom Order," and how Christian believers should have a mind-set of "Kingdom Dynamics" to relate to each other as joint co-laborers in Christ.

I am extremely grateful to my two sons, Jabutarik and Amad; who never gave up on believing that God was using me, while I was in my state of brokenness. I am also grateful to my mother, Willie Ruth; who heard my dream with a cheerful gleam in her eyes.

I would also like to thank my illustrator Paul Dollins, for his insightful illustrations in this book.

Additionally, I must express my gratitude for the support given to me by the Alfred family, through the offering of various resources.

At this time, I would be amiss if I did not give respect to Bishop Noel Jones, a Bible scholar to the highest degree, for encouraging me to "dig up my gift." Through his ministry, he has taught me how to be a spiritual critical thinker, and how not to be afraid to think outside of the box, so that I might maximize my experience in Christ.

Thanks are also in order for Dr. Larry J. Lloyd (Ph.D in theology), for his ministerial leadership teachings, counseling support, and most of all, his friendship.

Additional thanks are also in order for Bishop Howard A. Swancy, who provided essential godly influence through his tenacious preaching concerning the Oneness of God, and the teaching of the Apostles' doctrine.

Above all, with the highest praise, I give thanks to Father God for the privilege and the honor of being used by the Lord to write a book such as this. Without God, I can do nothing. But with God, I can do all things through Christ Jesus.

Introduction

This book is intended to be a spiritual and biological glossary; meant to dichotomize church activities, ministries, and positions. It is also intended to be used as a study aide to help the Body of Christ to better understand individual purposes, callings, and to visualize our interdependent positions within the Body of Christ. Upon reading this book, it is my hope that readers will begin to experience spiritual, psychological, personal, and physical wellness from a biblically based Christian perspective. Many of the terms used in this book are based upon my interpretation of the Bible, and my visionary perception of what I call the "Anatomy of the Spiritual Body."

Before I wrote this book, I experienced an epiphany that allowed me to grow from the strength of the knowledge contained therein. Furthermore, I was inspired by the Word of God to write such a book. It is primarily based upon the revelatory expression of what is known as the "New Order," after the prophetic pattern within the personification of Christ Jesus. In many ways, the writing within this book is referring to his Levitical order as it pertains to Melchisedec. Out from Him evolves the new Kingdom Order. Whereby, Jesus reigns as King of kings, and the Lord of lords. Through this Mosaic pattern, Jesus holds three offices as Prophet, Priest, and King! He has paved the way! He has paved the example for Christian living and service, through embodiment, action, and demonstration of the conduct by which we are to live, operate, and serve.

This is where the two patterns meet, and the earthly and heavenly rulership merge together to form a communion of both Heaven and

Earth! Thy Kingdom come, thy will be done on Earth, as it is in Heaven. Amen!

The following are references from Bible verses:

> *And it is yet far more evident: for that after the similitude of **Mel**chisedec there ariseth another priest* (Heb. 7:15)

> *Whither the forerunner is for us entered, even Jesus, made an high priest for ever after the order of **Mel**chisedec.* (Heb. 6:20)

Note: There are benefits to using some Christian religious jargon and Christian spiritual code words. This book utilizes some Christian spiritual code words and religious jargon terminology, just to state a point of reference for those who have some knowledge concerning biblical doctrinal belief. Religious biblical jargon serves as biblical code words or scriptural reference phrases that are communicated for the benefit of those who share and understand their meanings. Religious jargon stands to benefit those who share common knowledge in a belief system designed to quickly express common thoughts, but should not be limited to that thought, for the sole opportunity to grow or excel to another level of maturity. Included are some additional newly formulated terms I call "Christian Spiritual Code Words." The "code words" are terms that I have coined throughout this book. However, for the benefit of others who may not fully understand, the specific concepts and terms that will be discussed in this book, I have worked extremely hard to utilize descriptive terms and pictures to help convey the practical applications that are the most relevant to the reader.

Added note to the readers: Throughout this book, I will refer to the natural anatomy of the human body before introducing the spiritual aspects and functions of the Spiritual Body. If we can relate to the natural things that are merely types and shadows of God's truth, then we should be able to gain a more broadened perspective, and better understanding of spiritually inspired truth. By doing so, I believe that you will be better able to understand the terminology and functions of each element of the spiritual anatomy.

I firmly believe that if you read this book prayerfully, it will most certainly speak to you.

Anatomy of the Spiritual Body

As born-again believers, we have become a part of a spiritual matrix. We are fused together within the Body of Christ. This Body is not fleshly, but is instead a spiritual body of believers that's unified in Christ through being baptized into the Body of Christ, and filled with the Holy Spirit. So as born-again Christians, we have been "bought" with the price of Christ's Blood, therefore enabling us to become members of the Body of believers. The anatomy of the Spiritual Body is very much similar to that of the physical body. In many ways, living souls navigate systematically as celestial beings in a fashion similar to the solar system. Spiritually, living souls are inherently aware and in agreement with the Creator as a point of contact towards circumspect positioning. However, rebellion and ungodly influences can cause a disconnect to God. The desire to reconcile back to God, can occur through consequential circumstances that poses various possibilities of acquired happenstance, ultimately resulting in the convicted soul being led to repent. Repentance produces necessitous change in one's spiritual state by obtaining godly order through the benefits of God's guidance, direction, and realignment. God Almighty, is the One True Source and Master Architect of the entire universe. He's God all by Himself, and without need of any other. The Lord God has created and arranged all things according to His own determination, counsel, and pleasure. His severalty, sovereignty, omnificence, omnipresence, and omnipotence garner Him the right to be called God. I'll expound somewhat on a few of His attributes. **Severalty:** a quality and condition of being separate and distinct, having the right of possession, or ownership that is not shared with anyone. **Sovereign:**

one that exercises supreme, permanent authority and is self-governing. **Omniscient:** one having total knowledge. **Omnific:** all-creating. **Omnificence:** the state of being omniscient, having infinite knowledge. **Omniferous:** all-bearing; producing all kinds. **Omnipotent:** all-powerful, one having unlimited power or authority and all force. **Omnipresent:** present everywhere simultaneously. **Infinite:** having no boundaries or limits; also being immeasurably great or large, and boundless in all of His attributes. **Righteous:** morally upright, without guilt or sin, and morally justifiable. **Majestic:** belonging to or befitting a supreme ruler. **Holy:** It is God alone who is worthy of worship, veneration, in addition to our reverence.

*And I heard as it were the voice of a great multitude, and as the voice of many waters, and as the voice of mighty thunderings, saying, Alleluia: for the Lord God **omni**potent reigneth. (Rev. 19:6)*

*And ye shall be **holy** unto me: for I the Lord am **holy**, and have severed you from other people, that ye should be mine. (Lev. 20:26)*

Righteous *art thou, O Lord, and upright are thy judgments. (Ps. 119:137)*

God Effect

God is the Creator, Guardian Host, and Chief facilitator of both the spiritual and human bodies. Both bodies have relative comparisons.

Spiritual and physical similarities are a reflection of God being the Author and Creator. God is the Heavenly Father of His born-again offspring, and Husband of the Bride of Christ. He dwells in the Church Body of believers. The Spiritual Body is anatomically designed for perfection. And the craftsmanship of the physical body is evidence that such an awesome God does exist. He's also the Potentate and Orchestrator of both physical and spiritual anatomies. Being also the Alpha and Omega; outside of Him nothing exists. God is the effect of all things, and all things are created by Him.

For all things were created by him and for him. To God be the glory forever! Amen . . . (John 11:36)

God's Word goes out and performs whatever has been spoken; it's impossible for His Word to fail, and His Word will never return void of fulfillment. God is His Word. In fact, His Word became flesh in the embodiment of His Son, Jesus Christ our Lord. His Word is life, and we shall live by it. His Word is eternal and spiritual. The world and all of creation was formed by the spoken Word of God.

The Subject of Cause and Effect . . . God is the cause behind the creative effect, because God is, and so shall it be. The effectual working of God's Word creates, permeates, and controls whatever He says is so. The effect of God's Word alters physical matter, spiritual realms, and realities.

God's Word never dies or fades away. His Words generate abundance and it's a sustaining power. In actuality, His Words have a rippling effect which impacts everything it encounters. That which was spoken by HIM is forever settled in heaven, and will continually reverberate throughout the heavens and earth. Believers are to echo God's spoken Words throughout the earth by faith, and should be in agreement with its content. Members of Christ are called to be replenishing agents who work in order to produce a great exchange of God's innermost and limitless resources. We are to freely give as God so freely gives to us.

Believers should embrace God's Word, and abide therein for the purpose of Sanctification, which sets up apart for his purpose. Then, are believers reconciled and considered holy unto him. His Word signifies imputed righteous solidification, which shows forth each child of God to be justified by faith. Having the indwelling of the Holy Spirit is an assurance of being chosen by God. Having the Seal of God's approval will connect you to your true self in Christ Jesus. Believers are able to discover their identity in Christ Jesus. It's also more proof that you're authentically children of God. We are to be endowed with the Holy Spirit through the Word, along with His Delegated Power, which enables us to live. This is an awesome spiritual exchange of having allowance and access, with authority in stewardship. Being strengthened, gives members of the Body of Christ the mobility of expression and life, towards exercising our destiny in him (gifts and callings). God's Word is a beacon of light unto the world, casting off darkness and bearing witness to the truth. Through the Word, the Body of Belivers are given the ability to connect with Him as the ultimate source of fulfillment for our every need. The Lord makes us to be completely whole, through the Word. The Lord makes us to be completely whole, lacking nothing. The Word allows believers to partake of His Wisdom and Knowledge, which enables us to be a tool for the Lord God's Handiwork. Additionally, we also become co-laborers in Christ. Through faith and obedience in the Word of God, we are prepared to perform every good work. The Word provides us with Protective Garments that not only shield us, but enable us to build up the resiliency needed to work and battle within the spiritual realm.

Father God's Word causes believers to be faithful, fruitful, and steadfast in the truth, so that we might have dominion over the world's system.

Kingdom Dynamics

Kingdom Dynamics are the support system of the biblical Kingdom Order mind-set. This support system serves to build and advance the Kingdom of God. It is supported by God, and is the basis by which the Apostolic Order is established, therefore giving Christians dominion over the earth. Members within the Spiritual Body are to establish God's Orders here on earth, by way of His Divine Plan and Power. Following His Plan of action, believers are to become champion kingdom builders that advance God's reign, and establish His Authority here on earth. Kingdom Dynamics is a biblical mind-set that builds a systematic support system, so as to produce wholeness and strength within the Body of Christ.

Some of the advantages in following God's Kingdom Order:

The Body of Christ will operate in the kind of faith that advances life and godliness.

The Body of Believers will begin to operate within the Kingdom Dynamic mind-set that produces synergy and cohesion within the Body. Through this mind-set, members also serve a source of strength from one to the other.

Taken further, individual and collective worship, praise, and love will elevate believers to a higher level within the Body of Christ.

Amongst believers there will be no need. Believers will be completely whole, lacking nothing.

Now, that believers know their purpose and function within the Body, they will begin to assume their rightful place(s) therein. Furthermore, they will begin to take heed to individual and cooperative callings.

The Spiritual Body of believers, shall obtain its inheritance through Christ to manifest and show forth the promises of God.

Two of the primary objectives of the Kingdom Order are: to build Kingdom establishments and set godly order.

God's Order will be established throughout his Kingdom, and his foundational laws shall be proven to be righteous, holy, and perfect just as He is. Having been connected with the Lord through rebirth, believers should inherently exhibit the characteristic identity of the Heavenly Father.

Our foremost attention and reverence is then given to God, who stands as the Center of our Universe, and is glorified as such.

Of the increase of his government and peace there shall be no end, upon the throne of David, and upon his kingdom, to order it, and to establish it with judgment and with justice from henceforth even forever. The zeal of the Lord of hosts will perform this. (Isa. 9:7)

Mind=set and Purpose

The purpose of the Church is to be a conduit of God's ordinances, and to establish precepts through His Word, which brings forth love, power, and life giving force. It is also the job of the Church to bring forth the message of Christ's resurrection, which constitutes God's plan for man's reconciliation, and the healing that was purchased for us through Christ's suffering. Whereby, establishing the godly order which brings forth the message of salvation to the lost, and helps to further unify all born again believers. The Spiritual Body is intended to be a **cohesive support system** (unified, organized, collaborative, interrelated, interdependent, solid, strong, and true). Each member of the Body of Christ should operate within the framework and guidelines of God's purposes and mandates, which are built upon the sure foundation of God's Word. I'm tempted to use the word if, but not so! "When" the Body cooperatively works together on one **accord** and in one mind, all while operating under divine authority, then will the world see a glorious Church! Other examples for being on one **accord** would be as follows: acting in agreement with what God's Word says is true, working together in harmony with one another, being in concurrence as one complete and holistic systematic unit, while being strong in the unity of the faith. When believers begin to heighten their abilities in spiritual critical thinking, and respond in love accordingly, then believers can become totally whole. Lacking in nothing. God's Will calls for believers to adhere to and live by the Word. By doing so, God will see to it that all the Body's needs are met. This means no more concerns about finances, physical well - being, and spiritual malnutrition. Moreover, believers are also protected from hurt, harm, or any other elements of danger. That's because the Heavenly Father has already set in order spiritual and physical provisions to fulfill our every need. However, the Spiritual Body must align itself with God's Order to be blessed beyond measure. Believers should no longer need to depend upon this present worldly economic system. Through the Kingdom Order of spiritually aligned Kingdom Dynamic principles, which are revealed through scripture and emphasized throughout this book, readers will be able to discover the untapped resources that are available through Christ. These are the very resources that work to produce spiritual and natural wholeness. In Psalms (84:11), God tells us that He will withhold no good thing from those of us who walk uprightly. In fact, Christians already have the means to produce their own banking system, which is based upon an ethical code of merits. There is a Spiritual Kingdom that is set under the guidelines of God's Kingdom Order. Within the Spiritual Kingdom are applications of service to the King above all kings. There comes also the appropriations for services and wholistic well being. One clause is that you cannot serve two masters. Either you will love one system or hate the opposing system. Now, for those who think that this system is primarily for financial gain; it is not. You are automatically out of godly order if your motivation is driven by greed, pride, hatred, or deceit. I will touch on these points later in this book, but for now, I will describe some financial benefits in the Kingdom. If we are properly aligned with Jesus spiritually; having Him stand as the CEO and Kingship over all financial matters, the the Father will see to it that we have an unlimited surplus of spiritual and natural riches. This increase is not to be used for your benefit alone, but it is also to be used towards the benefit of all within Christ, and stands as an added resource to serve others who are in need. The family of God should acknowledge our Heavenly Father as being Provider, Protector, and Divine influence. Believers need to get in line with God's directives, and humble ourselves, then will all things align properly. Your response will determine your result . . . React in a godly fashion to receive a godly outcome that's blessed! Your actions should include some or all of these modalities to acquire spiritual success and enrichment, through prayer, reading the bible, fellowship, and meditation upon God's word. There should also be time allocated for fasting. These principles, aided by righteous motivation towards action and the following of ethically sound doctrine will further enhance the lives of believers. That which is tried and true will maintain its integrity.

> *Fulfill ye my joy, that ye be likeminded, having the same love, being of* ***one*** ***accord****, of one mind.* (Phil. 2:2)

*And when the day of Pentecost was fully come, they were all with **one accord** in one place.* (Acts 2:1)

*And they, continuing daily with **one accord** in the temple, and breaking bread from house to house, did eat their meat with gladness and singleness of heart,* (Acts 2:46)

*And by the hands of the apostles were many signs and wonders wrought among the people; (and they were all with **one accord** in Solomon's porch.* (Acts 5:12)

Wealth and Riches within the Kingdom

Wealth and riches within the Kingdom of God do not limit themselves to material and monetary gain. There are spiritual riches and health benefits as well. Dont allow the love of money to become the God of your life. It won't really profit those under its enticement anyway. For the Bible reveals that the Kingdom of God is more than meat or drink; our real treasure is the joy and peace that can be found within the Holy Ghost. The Kingdom of God is within our midst. In fact, the Kingdom of God is within those who believe. In God's presence there is peace, liberty, and fullness of joy. Christians shortchange themselves by not allowing faith to usher them into the Spiritual Rest that God has made available for every true believer within Christ Jesus.

The work of Kingdom Building does not measure success according to financial gain. However, it is one of the benefits and working tools of the Kingdom-building agenda. There is a biblical concept to obtaining financial leverage toward Kingdom advancement: You must remember that the Kingdom wealth system is not solely for the private use of any one person or church ministry. This kind of wealth is meant to be circulated and shared as a result of corporate collaborative efforts among the entire Body of Christ. Paying tithes and offerings are necessary and should continue as an act of one's faith and obedience. However, there are additional ways for supporting and sowing seeds. You may also consider partnering and planting into our incorporated global ministry, which is linked up with other believers, for a shared harvest return of wealth and benefits. Currently, the world is in the grips of a devastating global financial crisis. The failing economy has exacerbated the failure of the banking system. Much of this overwhelming debt and degradation comes from greed, selfishness, and the lack of moral integrity. Overwhelmed with toxic assets and other debts, many banks are faced with the real possibility of becoming insolvent. Moreover, as more and more banks teeter on the brink of failure, reality shows us that government bailouts can't save them all. These failing banks are facing the possibility of becoming nationalized. Bottom line: taxpayers will end up paying, no matter what, and there is a very good possibility that all of the money that was lost may never be recovered. However, inside the Body of Christ is the holistic resolution to all of our needs. As each member develops in the principles of Kingdom Dynamics, they will become more spiritually connected and form holy alliances. These joined forces ought not to have any private agendas among each other, but should be solely committed to the betterment of each other, as a collective and cooperative part of one's self. Then, we will see that within the Church body lies all the resources that we may have need of. Although, the world is experiencing a global economic downfall, believers have the answer towards replenishing and recirculating their wealth, right within their grasp. Christians can structure their very own private banking systems, complete with private equities that will work to produce its very own economic stimulus program. For more information about this economic growth system, please visit our web site, www.KingdomOrder.net.

I must reiterate that obtaining financial riches ought not to be a primary objective of one's life. The expansion of God's vision does not solely consist of getting a larger building (house) or obtaining property. God provides us with these things to help expand the capacity by which we are able to work in his service. This is done by following biblically sound principles and applications. It's not the physical building that defines the success of a church, but it is instead the ability of the church to promote spiritual change, as well as influence the hearts and minds of those who would believe!

How do we go about promoting godly change? The Body of believers must present a quality of life that is desirable, adaptable, and sustainable. Furthermore, Christians need to show the world that their faith in the Gospel is unmovable. Their lives should reflect the mandates set forth by the Word. In doing so, the world will see that the Gospel is not about fulfilling selfish motives. Instead, it is about changing lives for the better.

The Gospel is the good news concerning the Kingdom of God, and the salvation that is acquired through Jesus Christ's saving grace. It deters us from having to live a life of eternal damnation, destruction, and judgment. Additional benefits in salvation come in the form of being spiritually and physically prosperous. Our quality of life and living should rise to an entirely different level, as we submit to the Lord's Will for our lives.

One way we can surrender to God's directives, would be to begin to a daily process of submitting to God's precepts, as directed by the Word and Holy Ghost. Moreover, we should continually strive to endure sound doctrine until the end. This process begins as we receive God's Love, Redemption, and Reconciliation!

Galatians 6:7 Be **not** deceived; **God is not mocked**: for whatsoever a man soweth, that shall he also reap.

What Kind of Harvest Do You Expect?

Reaping and sowing for your harvest: Whatever seed you sow will determine the type of fruit you bear. The amount or level of sacrifice that one plants will determine the level of abundance in your return harvest. You must learn to sow specific seeds that will produce a well-earned harvest ground at a seasonably predetermined planting time. Thereby, being able to determine what type of reaping is gathered in due season. This is a principle that affects both the natural and spiritual realms of manifestation. Believers are not to sow blindly, but are directed to initiate and perceive an expected end. Through these applied principles, Kingdom builders can multiply storehouses, both shared and individually gained. Again, I will mention that the purpose for unification is not just to obtain wealth, but to show forth the likeness of: God's love, wholistic balance, truth, righteousness, holiness, power, provision, peace, joy, beauty, and glory!

These investment principles are based upon a school of thought introduced by the Bible. This was a common way of thinking, as referenced in within book of Acts, (Chapter 2, Verse 44). which enabled Christians to share in the support of each other's burdens. Nothing cultic about it! Never should this be considered as a false religious sect full of extremist views, and unconventional lifestyles, as directed by an alluring leader. However, because of escalating need and uncertainty, the world has experienced a substantial paradigm that underscores the power and validity of the Kingdom Order.

*And they, continuing daily with **one accord** in the temple, and breaking bread from house to house, did eat their meat with gladness and singleness of heart*
(Acts 2:46)

Kingdom Order

Christian believers are fitly joined together into one Spiritual Body that is to be undivided in Christ. This Spiritual Body embodies a Spiritual Kingdom of dichotomized colonies full of interdependent and interconnected believers. The Kingdom is manifested outwardly through power in righteousness, along with joy and peace within the Holy Ghost. Believers, once they have been translated into the Kingdom, are now prepared to perform Kingdom work. They continually strive to please God, and uphold a certain level of respect among other spiritual counterparts. Members in Christ should consider each other as an extension of themselves and God. Also, members must come to terms with being a royal priesthood, peculiar people, and being major parts of God's plans and purpose. God's rulership over born again believers is set by His Order of directives; by way of God's precepts and principles. His Kingdom Order is established within the framework of a divine apostolic government. Kingdom service is under the authority of our King, God Almighty. Whereby, God's people are different pieces to the same mosaic pattern, and yet, are all equally necessary pieces to fit together as a whole spiritual fabric.

Spiritual development enables believers to advance in godly maturity, while demonstrating divine characteristics. Being submitted to serve as a conduit for the Triune Godhead; whose headquarters is in Heaven.

A godly mind-set ought to be from a Kingdom Order perspective. Carrying out righteous actions of inspired vision and instruction, to build the Kingdom of God. A description of some of the works in progress are to equip young men who dream dreams, old men who have visions, and daughters who prophesy (or speak out to give birth to God's vision for the future). This is done by offering them a blueprint of God's divine and all - inclusive architectural design plans. This ministry of reconciliation will show others the glory that is being released through their dreams and visions. According to the Gospel that Jesus Christ taught, believers are to preach, pray, and live the Kingdom of God. Teachers should preach about the Kingdom, and encourage conduct that ushers in the fullness of God's Kingdom. Christians are to actively build a relative example of the Kingdom that is in Heaven . . . Oh Lord, your Kingdom come and purpose be done on earth as it is in heaven. Even so, Come Lord Jesus!

[The Lord's Promise of His Spirit] Then, after doing all those things, I will pour out my Spirit upon all people. Your sons and daughters will prophesy. Your old men will dream dreams, and your young men will see visions. (Joel 2:28)

Kingdom Builders are to be prepared for any mission as armed champions of faith. Being directly linked to God, and connected with each member of the Body of Christ in cohesiveness for the advancement of the Kingdom of God. Fully persuaded in love, to help edify, support, and bear each other's burdens. When one part of the Body is weak or in need of support of **any kind, Those** who are able to assist and protect should move appropriately with conviction and sacrificial love to help restore, strengthen, and uplift any affiliated member of the Body. Assistance could come in the form of selfless giving such as: prayer, encouragement, finances, sharing, exchange of resources, information, protection, work, ministry, along with any other act of love. These acts are manifested to help benefit the establishing the Kingdom Order.

Then shall they also answer, saying, Lord, when saw we thee hungry, or athirst, or a stranger, or naked, or sick, or in prison, and did not minister unto thee? Then shall he answer them, saying, Verily I say unto you, Inasmuch as ye did it not unto one of these least, ye did it not unto me. (Matt. 25:44, 45)

There are present and future visitations of the Kingdom that comes in salvation and judgment. God will establish His rule over all kingdoms. Whereby, God's heavenly authority and established governmental ruler ship is here on earth, and will invade the kingdom of Satan. The present visitation of the Kingdom of God, is the redemptive rule of God. The Kingdom of God, in conjunction with the Spirit of Christ, will defeat Satan and all the powers of evil. Thus, delivering all souls from the powers of evil. God's mighty Kingdom work is delivering souls from all evil. Ushering in the future visitation of the Kingdom of God's rule through Christ, is established in the Kingdom Order that is the basis for the transformation of the material (world) order, through the event of rebirth. The future visitation of God's Kingdom will come at the end of the age (Matt. 13:36-43).

Purpose of the Spiritual Anatomy

The purpose of each member of the Body of Christ is to be a reflection of the image of God, and to be an animated expression of His Word. We are to love, honor, and glorify God. The Body is to present itself as the Bride of Christ without spot, wrinkle, or blemish. Presenting a quality of being chaste and adorned to perfection. The Bride of Christ is to enter a marriage covenant in total union with the Creator. The Lord is Husband over the Spiritual Body. This Spiritual Body is to exhibit three different offices: the Bride, the Church, and the Kingdom. As the Bride, we are the object of God's affection and correction. As the Church, we function to minister to God through worship from a priestly posture. A global objective is to preach the Kingdom Gospel to all that will hear, and to evangelize throughout the world. In the office as Kingdom; God is KING over all . . . His divine order is established here on earth, as it is in heaven. Believers are to edify each other in the faith, and operate with in the power of God as One New Man in battle against principalities and spiritual wickedness. Although, we are many sons and daughters of the Most High God, we unite together as one to conquer back what belongs to the Kingdom. The Spirit filled Church Body has the Power to overtake heathen kingdoms and dismantle their thrones. First and foremost, the Body must yield itself to honor God's position. Therefore, we are to fully acknowledge Father God as Ruler, and submit to His Absolute Power and Dominion. His total power and dominion. Give Him the glory! Your Kingdom come, Your Will be done! Before the Lord comes back for the Bride of Christ, there will be a great transformation at the end of all preparation in due process. As the Bride awaits the Groom, all things are set in Order to be received unto His Glory for the Marriage Supper of the Lamb.

Revelation 19:9 And he saith unto me, Write, Blessed are they which are called unto the **marriage supper** of the **Lamb**. And he saith unto me, These are the true sayings of God.

And he said unto them, When ye pray, say, Our Father which art in heaven, Hallowed be thy name. Thy kingdom come. Thy will be done, as in heaven, so in earth. (Luke 11:2)

The Spiritual Body of Christ must work to preform the Will of God in the earth. Being in agreement with the Laws of God's divine Order; through active involvement.

And the kingdom and the dominion and the greatness of the kingdoms under the whole heaven shall be given to the people of the saints of the Most High; their kingdom shall be an everlasting kingdom, and all dominions shall serve and obey them. (Dan. 7:27)

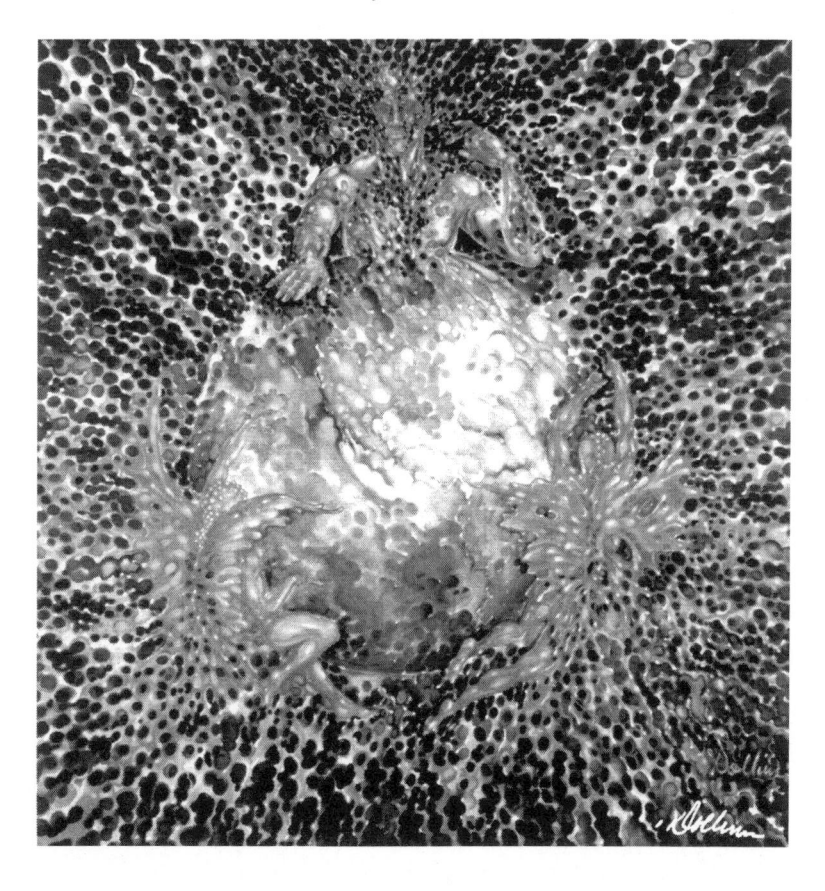

What Does God Look Like?

Our God is multifaceted exponentially, and is not limited to anyone's perception of finite understanding. And what does the Spiritual Body look like? Well, this description is not limited to any one interpretation. However, I can offer you a visionary glimpse of some of its expressions, interrelated functions, and value. This is an overview, but not limited to any of my mentioned functions and comparable relative perspectives concerning the Church Body. And, I must mention that if we were to all collectively imagine the magnanimity and the gloriousness of how our Lord has indeed fashioned His Bride the Church; we could not grasp the totality and the splendor of it all!

So, I will attempt to paint a vague visionary picture of the Anatomy of the Spiritual Body.

The Spiritual Body has spiritual senses that are able to see the presence of God, hear His voice, touch His face, and inhale His essence. "Oh, taste and see that the Lord is good!"

However, no earthly person has naturally seen God's glorious Face and yet lived, because God is a Spirit, and those who worship Him must worship Him in spirit and in truth. Yet, believers have beheld the Face of Him through the image of His dear Son, Jesus Christ. The Bible tells us that if we have seen Jesus Christ, we have seen the Father. Jesus is the fullness of the Godhead Bodily. And since Jesus has ascended back into heaven; He sends His Holy Spirit to remain here with us in His stead as our Comforter and Help. Now it's up to God's people to be that living embodiment of His expressed image on earth. And we are to be a reflection onto the world. The Body of Christ believers are to transmute the image of God as a collaborative bodily expression through the united efforts of an interconnected Spiritual Body in Christ.

God saved us so that we might be conformed into the image of His Son, so that we might become more like Him.

Our God is a triune God. He is Father, Son, and the Holy Ghost. And all these three are One! Jesus Christ is the living embodiment of the fullness of God. Jesus Christ was born of the Father; He came to us in the name of the Father, served as the Sacrificial Lamb to a world of sin, while also reascending to heaven to sit at the right hand of the Father in our behalf. Moreover, He sent forth the Holy Spirit (Comforter), to all who believe in Him.

The people of God are made in the expressed image of His Dear Son. Having been given the same triune nature of the Heavenly Father. This triune nature is made up of body, soul, and spirit. A special entitlement of the salvation experience, is the evidence of being born again within the Body of Christ, by way of receiving a newly circumcised heart. However, it is the responsibility of each believer to put on a new mind-set that is transformed, through the regenerated affects of God's Word, and applying it, by putting on the Mind of Christ. An additional benefit comes in the form of having a new life that is renewed through the born again experience of receiving the Holy Spirit. This empowers each believer to become all the Lord has intended for His offspring to be through Christ. This Spiritual Body obtains the faith to achieve victory, by overcoming every obstacle through life in Jesus Christ. This is done by dying to sin, and exchanging mortality for immortality.

Names and Titles of God

The Lord reveals Himself through various encounters with people. God is known to be the great Facilitator, and Supplier of all necessities, regardless of what the circumstances may be. His names reveal descriptive qualities of His infinitely faultless characteristics. He is the personal God, who meets us at the very point of our individual needs. In the Old Testament, you can trace God through more than 30 different names and titles in which, He manifests Himself to address particular situations that only He can handle. But in the New Testament, you will find only one name that is above every other name to be saved, and which to live by. The Name of Jesus is the only one used to declare the oneness of the Father, Son, and Holy Ghost. There have been great controversies over the name of Jesus, and what it represents. When we embrace the Name Jesus; then we embrace all of what salvation offers.

We can see Him to be the God who supplies all of our needs according to His riches and glory in Christ Jesus!

Here are names and references of God in the Bible:

S. No.	Name	Scripture	Meeting Point
1	Elohim	Genesis 1:1	God
2	YHWH oryh	Genesis 2:4	Lord God
3	El	Genesis 14:18	Most High God

4	Elyon	Genesis 14:18	I Am The Lord
5	YHWH (Yahweh)	Genesis 15:2	O' Sovereign Lord
6	Adonai	Genesis 15:2	God Of Highest Heaven
7	El-Roiy	Genesis 16:13	God Is Seeing
8	El-Shaddai	Genesis 17:1	I Am The Lord God Almighty
9	El-Olam	Genesis 21:13	The Everlasting God
10	Jehovah-Jireh	Genesis 22:14	The Lord Will Provide
11	Eheyeh asher Eheyeh	Exodus 3:14	I Am That I Am
12	Eheyeh	Exodus 3:14	I Am
13	YHWH	Exodus 6:3	God Almighty
14	Jehovah	Exodus 6:3	God
15	Jehovah-Rapha	Exodus 15:26	The Lord That Heals
16	Jehovah-Nissi	Exodus 17:15	The Lord Our Banner
17	Jehovah-M'Kaddesh	Exodus 31:13	The Lord That Sanctifies
18	Adon	Joshua 3:11	Lord
19	Jehovah-Shalom	Judges 6:24	The Lord Our Peace
20	Jehovah-Saboath	1 Samuel 1:3	The Lord Of Hosts
21	Eloa	Nehemiah 9:17	God
22	Jehovah-Elyon	Psalm 7:17	The Lord Most High
23	Jehovah-Raah	Psalm 23:1	The Lord Is My Shepherd
24	YH (Yah)	Psalm 68:4	God
25	Jehovah-Hoseenu	Psalm 95:6	The Lord Our Maker
26	Wonderful	Isaiah 9:6	Son Is Given
27	Yeshua (Jesus) Hebrew	Isaiah 42:12	Salvation
28	Jehovah-Tsidkenu	Jeremiah 23:6	The Lord Our Righteousness
29	Jehovah-Shammah	Ezekiel 48:35	The Lord Is Present
30	Elah (Aramaic form)	Daniel 2:18	God Of Heaven

31	Messiah (Mashiah)	Daniel 9:25	Deliverer of Israel
32	Jesus (Savior) ("Yehoshua" in Hebrew)	Matthew 1:21	Lord Is Salvation
33	Emmanuel	Matthew 1:23	God With Us
34	Jesus (Yeshua) (Mesi'-ah)	Romans 15:13	God Of Hope

Jesus is the embodiment of the fullness of God. When you call on the name of Jesus, you stir up the presence of God, and all His Grace will measure. He is the personification of all the names mentioned; combined in One. When you say Jesus, you've said it all!

Spiritual Body

The ***Spiritual Body*** is referenced by many names, The Body of Christ, Christians (believers), the Church body, the Bride of Christ, the Church, the Temple, Abraham's bosom, and the Kingdom. It is a Spiritual Body comprised of Blood-purchased believers, by way of the Redemptive work through Jesus Christ. Without God's plan and finished work of redemption, there would be no Church Body of believers. What is the Spiritual Body? It is the Spiritual Anatomy of believers that are baptized into the Body of Christ. The Body has many members and diversities of gifts operating in one Spiritual Body. This is an infusion of born-again believers that are baptized into the Body of Christ, through the Holy Spirit. What is the anatomy of the Spiritual Body? Describing first, the spiritual anatomical part of the skeletal sphere which is an outer spiritual covering that is similar to the physical skeletal frame. This covering protects and supports soft organs, tissues, and other parts of the vertebrae. In the same way, does the skeletal sphere of the spiritual body works to cover, protect, and hold all the inward spiritual parts together. The Skeletal Sphere is the protective work of the Holy Spirit. His presence guards and protects the Spiritual Body from falling or dividing amongst itself. Moreover, the Skeletal Sphere also fortifies the body against any satanic or adverse attacks from harboring within the Body of Believers.

So we, who are many, are one body in Christ, and severally members one of another. And having gifts differing according to the grace that was given to us, whether prophecy, let us prophesy according to the proportion of our faith;

or ministry, let us give ourselves to our ministry; or he that teacheth, to his teaching; or he that exhorteth, to his exhorting: he that giveth, let him do it with liberality; he that ruleth, with diligence; he that showeth mercy, with cheerfulness. (Rom. 12:5-8)

The Denominational Divide Dilemma

Here's an analysis to the denominational divide argument. Otherwise known as: "the big religious argument." Infuriating questions particularly among Christian denominations; such as: Who is this Jesus the Christ, and how did He come into existence? Was He the embodiment of the Triune God, and the total fulfillment of all that is, was, and ever will be? When all Christian denominations arrive at the knowledge of His fulfillment, then all division will cease. All will be as one, even as God is One. With the exception of any non-Christian or anti-Christian religions. There will always be division between believers and non-believers. Only born again believers are joined as one within the Body of Christ. I have come to the conclusion, that as the Word of God continues to open my understanding to the fact that the name "Jesus" bridges all gaps and unifies the Church Body as one. Jesus is also Yeshua the Messiah, whom the Jews are still looking for to establish justice in the earth (Isaiah 42:4). Actually, no one can come to the revelation of who Jesus is, unless the Holy Spirit enlightens the mind to perceive this truth. Those that attempt to understand without the Holy Spirit see this truth as foolishness, and cannot enter into the Kingdom of God. Without the Holy Spirit one would remain outside of Christ, and void of any real spiritual understanding. This is where you get religious people that are attempting to serve God in and of their own flesh. So, while others may joke about those who are oneness people or holiness people, the truth of the matter is, if they have the Holy Ghost and have been baptized in the name of Jesus Christ, then they still have a better chance of obtaining true fellowship in Christ.

So, it is important that we share with those within the various Christian persuasions the width, depth, and fullness of the knowledge of God, by introducing them to the Holy Spirit. And we must not abandon certain religious practices that help to revile the Word of God, and the Personification of Who He is. Without the Holy Spirit, those who are unbelieving and lacking faith, will remain in their sin. The doubtful and unbelieving will remain outside, looking at something that appears strange, but will nonetheless, continue to blindly follow after the flesh in a religious and undone condition. In this current dispensation of grace, there is no salvation obtained unless it comes by way of repentance. Furthermore, the act of repentance is aided by submission unto the Lord as Savior and God. In order to walk in the faith, the believer must die out to the deeds of the flesh, through baptism under the name and authority of Jesus Christ for the remissions of sins. Repentance first begins with awareness of sin through the presence of God's grace. Jesus Christ is the Righteous covering for a sinful life. Without this realization, attempting to serve God by merely doing good deeds is worthless. Making it impossible to walk in the fullness of life or enter into His Kingdom. Furthermore, when it's all said and done, God will say, "Depart from Me, I never knew you." So, walk circumspectly, according to the Lord's percepts and ordinances.

The Essence of Life

The Blood of Christ . . . The Living Water . . .

The Blood of Christ declares the Body of believers as being redeemed of the Lord, whereby His Blood is the outward and inward spiritual protection applied upon the Spiritual Anatomy of the Church Body. This outer layer acts as a spiritual protective sphere that's similar to the skin of the human body, which guards against outside elements. The spiritual covering is the Blood of Jesus Christ. The Blood of Jesus is the sacrificial offering that purchased the ransomed debt in full for every sinner held captive, and for remission of all sins. This is where the Blood of Jesus is accepted by grace, and received through faith in Christ through the preached gospel of Jesus. The redemptive Power of the Blood of Jesus enables the Body to come boldly to the throne of God. The sinner is washed clean through purification, and the life that is within the sanctifying Blood of Jesus. Christ stands as the Anointed One of God, who paid the price of redemption by being the Sacrificial Lamb who was slain for the sins of the world. Christ Jesus' shed Blood on Calvary's Cross cleanses us from of all unrighteousness of sin. Receiving the Lord's grace through faith, will allow the sinner to be reconciled to enter His care and protection. Without God's protection, the Body would be open to the elements and subject to invasions and attacks. The Blood is within the spiritual skeletal frame of the Spiritual Anatomy. And the skeletal frame is our armor of God. Life is in the redemptive Blood of Jesus, and His saving living Grace is in the Blood, through the Power of the Holy Spirit. In the Spiritual Body (Anatomy) of Christ, what I would refer to as the "*Life Stream*" flows both Spiritual Blood and Living Water, throughout the entire Spiritual Body of believers within the Body of Christ.

***Armor of God：** The Lord of host prepares the army of the Lord to resist the enemy, and endure against all troubles. To overcome any battle by faith in Jesus Christ, through the Power of the Holy Ghost; to obey His Will, to understand, and properly implement the principles of God's Word. So, put on the whole armor of God. Don't be caught dead without being clothed with the imputed righteousness of God by faith, and submit to Him as Lord over your life. Without this righteous suit of armor; it is impossible to resist the devil and win.*

> *For we wrestle not against flesh and blood, but against principalities, against powers, against the rulers of the darkness of this world, against spiritual wickedness in high places. Wherefore take unto you the whole armor of God that ye may be able to withstand in the evil day, and having done all, to stand.* (Eph. 6:12, 13)

The Living Water: This is the Life-giving force that comes from the power of the Holy Spirit. Power and redemption comes from the anointing of the Holy Spirit. Out from this spring of Living Water flows eternal and abundant life to believers in Christ. Believers are made pure and holy by God's imputed righteous purification through faith. The mind, body, soul, and spirit are spiritually renewed and transformed by the washing of the Word. Redemptive virtue restores the believer to right standing before God by the Holy Spirit through Christ Jesus. The Holy Spirit indwelling within you is your seal of divine favor and assurance before God.

Sanctification and righteousness is applied to each believer through the redemptive work of the Holy Spirit of God. The believer is then reckoned to be set apart from the world as children of God that are justified and holy. The Living Water nurtures and sustains the Body of Christ, and gives each member power to move in activity and expression. Also, the Spirit filled believer is given power to stand in adversity. This Living (Spiritual) Water flows within the Blood of Christ as the "***Life Stream***" throughout the Spiritual Body of Christ.

> *He that believeth on me, as the scripture hath said, from within him shall flow rivers of living water.* (John 7:38)

Human Comparison

Both Spiritual and physical bodies come complete with features such as the head, which is equipped with a brain, face, eyes, ears, nose, jaws, and a mouth. The neck connects the head and body together, and is attached to a spinal cord that connects the neck to the upper torso, shoulders, arms, wrists, and hands. The lower torso is connected to the waist, hips, legs, ankles, and feet. And yet all these parts, including the internal organs, work concurrently to function, and to complete specific tasks. Each member of the physical and Spiritual Body is reliant upon each other to perform at their maximum level of potential, as orchestrated by divine intervention. Even though we may know more about the physical body, we can learn more about the Spiritual Body, by simply observing the psychological aspects of the physical body. In doing so, we can get a better glimpse of the similarities between the two bodies. Throughout this book, I will make comparisons between the Spiritual and physical anatomies, and showcase how the Spiritual Anatomy (the Church Body of Believers), equates to physical anatomy of the human body. Throughout the Bible, you will find numerous comparisons to the natural body, as it relates to growth and the Spiritual Church Body.

For as the body is one, and hath many members, and all the members of that a one body, being many, are one body: so also is Christ. (1 Cor. 12:12)

In our natural body, the **skeletal system** is a hard structure of bones and cartilages that provides the frame for the body. Each skeletal system consists of the following vital components, the head, neck, spine, ribs, shoulders, arms, hands, hips, legs, and feet.

The ***Spiritual Skeletal System*** represents the capacity to build and execute movement, based upon the sure foundation, which has been built upon the Word of God. This works to establish the strength and framework of the entire Body of Christ. It's the spiritual connectivity that joins the entire Body together. The framework is fortified by the Wisdom, Knowledge, and Power of the Holy Spirit.

In the natural body there's a ***musculoskeletal system***, which is composed of: muscles, tendons, ligaments, bones, joints and associated tissues. This enables movement of the body, and provides support to maintain stabilized form. The Church Body has a ***Spiritual Musculoskeletal System*** that is systematically similar to the natural body, yet, its framework is supernaturally structured to comply with the Will and Purpose of God. Through the efficacious Blood of Jesus and the Power of the Holy Ghost, the spiritual molecular frame has the capacity to resist the opposing forces of the enemy. The Power of God is endowed within the Frame of this glorified spiritual Body of believers. The Spiritual Anatomy of the Church Body of Christ is adored with, grace, righteousness, purity, peace, power, promise and truth!

Human Comparison

The anatomy of the ***human body*** is comprised of the head, which is the uppermost part of the vertebrae. It contains the brain, face, eyes, ears, nose, mouth, and the jaw. The jaw is in a separate compartment, which is referred to as the skull. The brain is that part of the central nervous system that includes all of the higher nervous centers enclosed within the skull. It's continuous with the spinal cord that's composed of nervous tissue containing cell bodies, as well as fibers. These fibers work together to form the cerebral cortex that consists of neurons. The brain is also the seat of consciousness, thought, memory, and emotion; more commonly known as the mind. Human consciousness originates from God, but is censored and manipulated through the brain, and is manifested especially in all facets of thought. Clinical studies of the mind conclude that "the mind is the principle of intelligence." The spirit of consciousness is regarded as an aspect of reality. Moreover, the mind is still the faculty of thinking, reasoning, and application of knowledge. However, ***Spiritual Consciousness*** originates from and remains pure through God. Since God gives mankind the entitlement of free will. His selfless Love, allows freedom of choice. But it is crucially important to consider His directional influence in all choices and decision making. This allows us to have a renewed mind through Christ, who is able to comprehend, adjust, and align to the principle concepts within spiritual critical thinking. Furthermore, this is all necessary to operate in the godly values that constitutes the statutes that make up God's Kingdom Order. Therefore, the human mind should carry collective conscious thought, and unconscious processes of sentiment that direct and influence all facets of human behavior (mental and physical). But godly consciousness is not based on, nor is influenced by, outside sensory devises. God consciousness remains unhindered and untarnished by worldly influences, and recollects that which is innately and inherently familiar. Both the spiritual and physical brains direct the body to operate within mobility. It is the primary center for the regulation and control of bodily activities; receiving and interpreting sensory impulses, along with the transmitting of information to the muscles and bodily organs. But the **Spiritual Brain** receives its charges and impulses directly from the Mind of God only. The God Source is infinite in wisdom, knowledge, understanding, and infallible truth. The Mind of Christ is never disconnected or independent from its one true Source, the Creator. Unlike the human brain, there is no inferior matter to cloud the renewed mind through Christ; it is at its highest state of consciousness. In the spiritual body, the spiritual brain is the Word of God, and from His Word flows spiritual perceptions through the mind of Christ. The spiritual link or pathway to the Mind of God is: God Himself, who has provided a Mediator in Jesus Christ, who provides a divine connection that transforms the mind of every believer. Jesus Christ is also the **Spiritual Head** of the entire Church Body of believers.

    Amen . . . God's covenant believers are to follow the leading of the Lord, by adhering to statutes as directed by following the Mind of Christ.

For who hath known the mind of the Lord, that he may instruct him?
But we have the mind of Christ. (1 Cor. 2:16)

Human Comparison

Human Anatomy—Receptor:

1. A nerve ending or other structure in the body, such as a photoreceptor, which is specialized to sense or receive stimuli. Skin receptors respond to stimuli such as touch and pressure,

which signal the brain by activating portions of the nervous system. Receptors in the nose detect the presence of certain chemicals, which leads to the perception of odor.
2. A structure or site found on the surface of a cell or within a cell, that can bind to a hormone, antigen, or other chemical substance. This works to initiate a change within the cell. For example, a mast cell within the body encounters an allergen; specialized receptors on the mast cell then bind to the allergen, resulting in the release of histamine by the mast cell. The histamine then binds to histamine receptors in other cells of the body, which initiate the response known as inflammation, along with other responses. In this way, the symptoms of an allergic reaction are produced. Antihistaminic drugs work by preventing the binding of histamine to histamine receptors.
3. *Physiology*—A specialized cell or group of nerve endings that responds to sensory stimuli.
4. *Biochemistry*—A molecular structure or site on the surface or interior of a cell, that binds with substances such as hormones, antigens, drugs, or neurotransmitters.

Spiritual Receptors: come from the indwelling of the Holy Spirit, as He administrates within the office of the Church Body of believers, and branches out from within the Spiritual Source of the Word and Life of God. These are spiritual nerve endings that receive godly impulses of instructions, spiritual vitality and life. Spiritual perception is a derivative to having Spiritual Receptors that are detected by way of the Holy Spirit. Spiritual Receptors translate spiritually transmitted impulses, and then discerns the origin of spiritual impulses regarding all commanded instructions. Whereby, determining the origin of every direct and indirect impulse. The Holy Spirit articulates to the spiritual reception; that which is true and holy. Through spiritual discernment, spiritual receptors seek to connect with the Holy Spirit to discern by way of spiritual wisdom, knowledge, and understanding. Spiritual receptors work in harmony with the Holy Spirit to obtain spiritual intellect; in order to decipher whether or not an impluse should be considered as direct contact with the Lord God. Spiritual receptors operate directly within the referential and justification guidelines of the Holy Spirit, in order to execute a righteous reaction. Spiritual receptors are also provided by way of the Holy Spirit a as warning system that recognizes the invasion of outside demonic and satanic influences. By way of God's Word, spiritual receptors are fortified with Truth to reject any ungodly impulses that would oppose the Knowledge and Will of God. Spiritual Receptors operate under God's spiritual laws to transmit and translate spiritual information throughout the Spiritual Anatomy in Christ. Each Spiritual Receptor holds messages of information that communicates the words and utterances of Jesus Christ. Spiritual Receptors also enable each believer to communicate to God through worship and in prayer. Their purposes are to receive God-given information through the Spiritual Senses of sight, taste, smell, touch, and hearing, in order to relay messages to Spiritual Neurotransmitters throughout the Spiritual Nervous System. These Spiritual Receptors initiate responses designed to produce the bodily actions needed to move forward a task or mission. Also, their responses are to ward off any life-threatening or evil devices, along with sounding off the alarm that notifies the rest of the Body when conditions are not right. These Spiritual Receptors, when threatened, assume a position for defense against any common threats within or outside the Spiritual Anatomy. Each born-again believer that remains connected to the God Source, Jesus Christ, is stimulated by Spiritual Receptors. At the moment a Christian believer is endowed with the Holy Ghost, they are then enabled to receive Power, provisions, and entitlement to the promised inheritance of covenant blessings and grace. Believers are baptized into the Body of Christ, then are planted within Christ to be made alive in Him. As Children of God, we then bear the fruit that is made from the branches of the True Vine. A divine connection with God, enables the believer to receive benefits such as: true Love, Power, Wisdom, Knowledge, Understanding, Restoration, Productivity, Wholeness, Spiritual Riches, Direction, Focus, and Joy. Spiritual Receptors receive God's instructions and fellowship. That allows a bridged connection of communications to remain open. Through this divine reception, the ability to retain spiritual fellowship with God and other believers is maintained within the Spiritual Anatomy of Christ.

Human Anatomy—Neurotransmitter:

1. A chemical substance that is produced and secreted by a neuron, and is then diffused across a synapse to cause excitation or inhibition of another neuron. Acetylcholine, norepinephrine, dopamine, and serotonin are examples of neurotransmitters.
2. Any one of a number of chemicals that are used to transmit nerve signals across a synapse. They are sprayed from the end of the "upstream" nerve cell, and then absorbed by receptors in the "downstream" cell.
3. A neurotransmitter that is derived from tryptophan, instrumental in the administration of processes such as sleep, depression, memory, and other other neurological functions.

Spiritual Neurotransmitters: These are transmitted by way of the ***Holy Spirit.*** This gives us the ability to pray, communicate, and have fellowship with God, conjointly with other joined members within the Body of Christ. Within this spiritual networking system is the ability to both relay and exchange God-given information and messages among members within the Spiritual Anatomy. Through Spiritual Transmitters, there is also the ability to influence, excite, encourage, and strengthen other associated counterparts within the corporate Body of Christ.

Spiritual Neurotransmitters (*Leaders*) are also composed of spiritual leaders within spiritual cell groups (church group or member affiliates) that receive stimulating spiritual impulses or instructions. These impulses or instructions are to be carried out within the infrastructure of the Spiritual Anatomy, along with any other outward facets of Christian service. Another function of Spiritual Neurotransmitters would be to relay or retransmit God's instructions and information throughout the Church Body. In addition, they communicate these directives and commands for the purpose of setting controls for the enhancement of spiritual conditions, and climates. Their interworking is within the Spiritual Body, by way of the Holy Spirit. They also communicate conditions within the Body as infantry leaders. These Spiritual Neurotransmitters send indicators to suggest whether the Church Body should: rest, watch, pray, work, retain information, or go forth in battle against an opposing enemy. These Spiritual Neurotransmitters operate and oversee in a position of authority from God.

However, the Spiritual Anatomy should be on guard against spiritual pathogens, which are disease-causing agents that work as harmful transmitters. These agents produce other pathogenic carriers that form together in order to demise, polarize, debilitate, and destroy that which is spiritually healthy within the Spiritual Body.

The Mind Is an Amazing Creation

The mind is like a vortex of sensory-related matter of connected energy that's governed by a related force. Whether it is good or evil, depends upon the source of its connection and influence. If the mind is reconciled back to the likeness of Christ and renewed, it has then transcended to a superior quality. Now, that mind is governed primarily by the Word of God through the Power of the Holy Ghost.

Make use of your mind in a methodical, analytical, and spiritual manner. There should be growth towards a level of maturity, and there should be discipleship in the ways of Christ. Believers should have a holy discipleship mindset to serve as a faithful student, willingly following after the patterns and instructions of Jesus Christ. There are various truths that should be carefully considered through the Word of God, and by way of the Holy Spirit, our Teacher and Helper. We must never cease to continually learn about the things of God. This is because his Word is inexhaustible!

The Bible points out, that the wise of the world can sometimes bring a particular awareness to our thinking. I find that Christian believers do not always use the ability to reason concerning some matters. Instead, there's a tendency of being a little more mystical and ethereal about things. God has given us minds that analyze, deduct, and process thought. However, on many occasions, that ability is abandoned. Because, there is a tendency to think that it's forbidden by

God to do so. Even though, the Lord has equipped us with minds that are able to reason through the consensus of His Word. His Word shall lead and guide unto all truth. It should be understood that God reigns upon the just and unjust. The unregenerate or unsaved soul, who has the God - given ability to use logic and reasoning, but fails to do so, has subsequently ignored God's sovereignty over their innate intellect. This would cause their earthly wisdom to be considered foolish, when compared to spiritual wisdom, which is cultivated and inspired by God. The unregenerate mind does possess the ability to acquire knowledge or ascertain facts, but it does not have the capacity to properly align this factual information with godly purposes. Worldly (conventional) knowledge is a tainted or a distorted form of truth, that is devised to accomplish negative operations. This is because of self - centered assent or evil intent in the sight of God. However, Christians can learn and glean from their fact findings, and therefore, should reject information that is untrue and unholy. Believers should reprocess information through the Word of God, so that it glorifies Him in the process! Don't rely upon worldly wisdom or conventional knowledge to be your primary source for gathering factual spiritual information. God has given the believer an ability to acquire witty and inventive ideas! Start developing your thought processes to operate in the fullness of the knowledge of God. This will bring complete wholeness and total fulfillment of the mind, body, and spirit in Christ.

Being Spiritually Balanced

The Mind of Christ is a mind of focus, one that's purpose driven, and governed by the Holy Ghost. In this present time, people are ridiculed for their dogmatic beliefs toward godly conduct, and their spiritualization of perceptional thoughts. Spiritual-minded people are sometimes looked upon as being crazy by the world's standards, for considering the Holy Spirit as the final and higher authority to guide the mind spiritually in decision-making. Although, the Bible says that God's chosen people would be considered peculiar and part of a royal priesthood; believers should be forewarned that a godly disposition is not considered mainstream, or widely acceptable. Yet, such a disposition is very much set apart from the world. Believers are to have a godly mind that submits to its True Source for inspiration, guidance, intellect, and purpose. A God-centered mind is a mind that listens to, and obeys the Voice of God. This kind of mind does not have private agendas of its own. It does not serve two masters. It remains connected to its True Source; God, who is over all creation. This mind maintains its integrity to the Word of God through the Holy Spirit. How can we determine if we are attempting to rationalize, and justify our thinking through a carnal mind? And how does the renewed mind transform to align itself with the Mind of Christ? We must learn to self-examine our pattern of thinking to see if it lines up with the principles of God's Word, while also examining our motives for decision - making. The Holy Spirit will allow us to determine, by way of the Word of God, whether or not, our thoughts are pure, true, and righteous. This is according to the fruit of the Spirit, in which we bear fruit. Through the spirit, you will bear fruits such as love, joy, peace, long - suffering, gentleness, goodness, faith, and meekness. Living in the Spirit helps with nullifying ungodly deeds and desires. Then we can begin to align our pattern of thinking to coincide with the Mind of Christ. This is accomplished by putting on the Mind of Christ. A godly thinking mind does not rebel, divide, or against itself, or its Maker. It knows its true self, and is spiritually conscious within the Body of Christ. However, there are times when God's people in their individual walk or fellowship will veer off their spiritually chartered course. Veering off or bypassing the course of righteousness, is due to a lack of discipline, sin, unbelief, or lack of maturity in any particular area of God's word. This person must desire God's hand of correction upon his or her life, in order to regain the proper focus and necessary direction. However, if a person is unsaved, then it is impossible to please God, or to live a life of righteousness within the Spirit. Unless a person is born again of the water and of the Spirit, they're only capable of fulfilling the deeds of the flesh. If a pattern of thinking is contrary to the Will of God; the Holy Spirit is able to examine the nature of it, and correct the process of each erroneous pattern. The Holy Spirit will allow for proper recognition in the way that is true. Then, there is no excuse for remaining ignorant in regards to the qualities, values, and standards that validate living according to the Spirit. Therefore, there are no corresponding actions that justify the satisfaction of evil and lustful desires. It is impossible to live the life of sin when one is living within the Spirit. You cannot live in and out the Spirit at the same time. The evil deeds of the flesh are manifested as: adultery, fornication, uncleanness, lasciviousness, idolatry, witchcraft, hatred, variance, emulations, wrath, strife, sedition, heresies, envying, murder, drunkenness, reveling, or anything else that mimics these actions. Ultimately, these actions will be judged, exposed, and identified by the light of God's truth. A spiritually balanced mind is one that is transformed by the regeneration of the Word of God. This produces a life of spiritual maturity, stability, productivity, and prosperity. The spiritual mind is renewed daily in Christ Jesus.

There is a condition of the mind, in which the Body of Christian believers tend to battle with, and that is the imbalanced mental state of being ***spiritually bipolar. However, the condition of the vassal's heart will determine if they are entirely in the Body of Christ, or if their position is entirely based upon an outside observation, which shows that they are not truly saved.*** One must determine if a tormented vassal is demonically possessed or oppressed. If the vassal turns out to be unsaved and under demonic possession, then they need the Word of God applied over their life through prayer and fasting for deliverance. Those who are called and equipped with power, can cast out any evil spirits in the name of Jesus Christ. After that tormented soul is set free, and the vassal is purged of demonic possession, then the freed soul must immediately accept the salvation of God, obey the plan of God, and receive the indwelling of His Holy Spirit.

A born-again believer, who is filled with the Holy Ghost, cannot be possessed by the devil, because one cannot be in the light and yet fellowship in darkness at the same time. One's spiritual house cannot serve two masters. As the bible declares: "You will love one, and hate the other." Those who may be undecided as to who they are serving, must ultimately decide this day, which master they will serve.

Christian believers can at times, experience oppressive thoughts from the enemy, suffer from emotional abuse, and even encounter spiritual abuse by the hands of someone considered to be a trusted friend within the church. Not only that, but some believers have been so seriously wounded and damaged by other alienating vassals, that they have developed severe mistrust issues. This is another ploy of the adversary to cause division, strife, alienation, sickness, and lack of productivity within the Body of Christ. Believers must counteract any evil strategic plan of strife and division, by remaining connected to God's Source of Life. A godly connection is enabled by submission to the Lord's Order as King and Chief Headship. God's Plan is that members commune together in love and fellowship with each other, and relentlessly serve in fellowship with Him.

My observations of spiritually wounded vassals lead me to append a term called "*Spiritually Bipolar.*" The only remedy for **Spiritual Bipolar** behavior is a proper intake of the Word of God. Then the Holy Spirit must provide illumination and clarity to ingest, metabolize, and assimilate the pure truth. It is also important to have the ability to eliminate what is false, by way of obtaining a healthy supply of the Bread of Life. Prayer should always be applied to counteract any form of illness within the Body of Christ. God's Word brings wholeness and spiritual balance. Jesus Christ is the Living Word, and the Bread of life. Spiritual members should live by every word that proceeds out of the Mouth of God. How should Christians deal with those who are struggling with greater degrees of mental illness, such as emotional or physical, chemical or hormonal disorders? Of course, believers know that with God, nothing is impossible. And God is the Divine Healer over all manner of sickness. Believers can lay hands on the sick, and by faith, believe the miraculous healing of God can follow with any cure. And all manner of healing and deliverance can prevail by the anointing through faith. Although, a vassal may become spiritually oppressed by an evil spirit does not mean that they're possessed, but simply demonstrates that they're currently under attack and oppressed by the devil's evil outside influences. If already saved, then that member can demand the enemy to depart. But until then, you can pray for the Hand of God to move, deliver, and set free. However, sometimes it may be necessary to receive medication from the doctor's care to aid in managing chemical imbalances, or to receive psychiatric help for emotional and behavioral problems. Even the knowledge to apply physical healing, or to correct certain wrongful behaviors, comes from God. But no one can physically heal what is spiritually induced. Only God can heal both physically and spiritually. God alone carries the remedy for the sin-sick soul.

And signs shall accompany those believing these things; in my name demons they shall cast out; with new tongues they shall speak; (Mark 16:17)

There is a difference between spiritual and conventional knowledge, and wisdom. Conventional knowledge and wisdom deals with that which is considered to be relevant and practical according to the world's point of view. Spiritual Knowledge and Wisdom knows what is practical, and is always relevant according to pure unwavering Truth from any position or stance. Pure Truth extends far beyond, to the unseen and hidden knowledge that is Eternally True; without wavering. It maintains validity, and remains unbalanced. There is an element of spiritual Alzheimer's disease among certain Christians; whereby, they will forget repeatedly what God's Word has truly spoken. They are unable to retain godly knowledge of the Lord's Wisdom. Also, there is the outsider who is unable to comprehend the things of God, as a result of being spiritually brain dead or spiritually retarded. This would be the individual that's not connected to the Mind of Christ, but operates instead merely by emotional and physical responses alone. And finally, there's a spiritual dysfunction among individuals that are spiritually retarded. Spiritual retardation can occur either at physical child birth or during continuous rejection of God's word. This dysfunction may develop as a result of their spiritual consciousness within being sheared. Those that may be spiritually separated by this void, are unrepentant, and live within a realm of spiritual darkness that does not all for comprehension of godly truth.

Another Form of Being Spiritually Bipolar

There needs to be balance. Otherwise, there's no real order. You can find yourself out of spiritual order, when there's religion without true biblical foundational doctrine. However, religious ordinances can instruct as a means of providing guidelines to practical methods, and can serve as a necessary action for a set pattern or guideline towards serving God properly. We can also identify spiritual disorder when there is a imbalanced level of spiritual cognizance. This spiritual disorder also highlights the consistent failure to align with godly discipleship practices. Properly aligning to the Word of God, brings spiritual balance that will enable the believer to more easily acquire knowledge and wisdom, while also performing actions that can lead to an overall better quality of spiritually healthy living. Sometimes, a vassal that is spiritually complacent or bipolar, can be introduced or exposed to some scripture and doctrinal belief concepts, but may still exhibit signs of immaturity and lack of overall growth. So, they resort to living double lives, in which they may appear to be a full part of the body, but are inwardly still wavering. The resolution for this type of behavior, is to become more deeply rooted in the biblical foundational truths that connect us to God's Word. It is in His Word that we find direction through His Wisdom, Knowledge and Guidance. It is this knowledge that helps us to attain the deliverance that promotes change. So, for the sake of being properly balanced in the things of God, we must commit to a certain level of religious discipline that's fortified by the proper understanding of God. This is done by establishing and maintaining a close relationship with God. The Lord is the Source of all the Grace that is needed for victorious living and growth. Experiencing this growth will lead unto the fullness of the knowledge of Him. If you walk or live in the Spirit, then you will exercise the fruit(s) of the spirit, and will not succumb to the lust(s) of the flesh. The fruits of the Spirit enable you to live a spiritually balanced life that is productive and prosperous. On the other hand, the lust of the flesh is a form of being spiritually deranged or crazy, and is contrary to the guiding percepts of God's plan. So, the mind must be renewed or regenerated through the Word of God. Therefore, we can develop a Christ - like mind set. As people think, so shall they become. The Lord would rather you be either cold or hot. He cannot stand for anyone to be lukewarm or not willing to live a Christ - centered life. In other words, He does not like counterfeits of any kind. So, you must be on guard not to embrace the world to the point that you become just like the unbelieving. Indeed, there may come a point to where there's no real distinction of which side you truly belong. Who you are as believers must be directly identifiable to God as the Heavenly Father. The Lord would prefer that your godly influences promote change in the world, and in the lives of those who would yet believe in Christ as Savior.

Questions that cause separation

When it comes to the questioning of moral ethics, I firmly believe that Believers should strive to adhere to that which is godly and righteous, even in spite of the potentially severe consequences.

Here's an assessment concerning a denominational divide argument. There are a few questions that lead to contention and division in the Church. The major questions of concern are: Who is this Jesus the Christ, and how did He come into existence? Was He the embodiment of the Triune God, and the total fulfillment of all that is, was, and ever will be? When all denominations arrive at the knowledge of His fulfillment, then the division will cease to exist. All will be as one, even as God is One. Here is the conclusion I have reached, as the Word of God has opened my understanding. The name "Jesus" bridges all gaps and unifies the Church body as one, through the Holy Spirit. And no person can come to this revelation except the Holy Spirit illuminates the mind to perceive this truth. Those that attempt to understand godly wisdom without godly knowledge that is attained through the Holy Spirit, see this truth as foolishness, and cannot enter into the Kingdom of God. This would then hamper their ability to have true communion with God. Furthermore, the ability to have real fellowship would also be hindered. This is where you get religious people that are attempting to serve God in and of their own flesh. This is equivalent to having a Pharisee - or Sadducee type of mind-set, in which the focus is placed on adhering to the customs of the law, instead of Christ. There is a form of godliness displayed, but the power thereof, is denied. So, while some may be considered outcast and ridiculed for being identified as oneness or holiness people. Yet, if these believers have the Holy Ghost, and are baptized in the name of Jesus Christ, they then have a better chance of obtaining true fellowship in Christ. So, it is important that we share with those within the various Christian persuasions, the width, depth, and fullness of who God is, and how to better understand His ways by introducing them to the Holy Spirit. Believers must not abandon certain religious practices that work to revive members by the Word of God. This would broaden your awareness in the Personification of who Yeshua Messiah (Jesus Christ) is. Without the drawing of the Holy Spirit that leads to repentance, individuals are not able to receive the urging of Christ, as He knocks on the door of the heart. The typical unsaved religious person would simply be looking at something that appears strange. Not seeing clearly, because they're attempting to follow after the deceitfulness of their own fleshly desires and motivations. Instead, they would only appear to be righteous, but on the other hand they remain superficially religious and undone. Salvation is not obtained, except through repentance and submission unto the Lord as Savior and God. The unsaved must die out to the deeds of the flesh, through baptism under the authority of Jesus Christ for the remission of sins. Then they must acknowledge their personal sin, repent, and then surrender their will over to God. Only then, can they receive Jesus Christ as the righteous covering for all sin. Henceforth, they can begin to walk in the fullness of life. Until then, they will only serve God through deeds only. Lacking total commitment to him. Upon judgement, God will say: "Depart from me, I never knew you."

Should There Be Division within the Body? "The Denominational Divide Question"

There shouldn't be any division within the Body of Christ. A kingdom cannot stand if it is divided. However, if necessary there should be correction, redirection, and undergirding. This may be due to language issues, interpretations, and misunderstandings.

This book was inspired by my growing concern of the lingering questions that lead to contention and strife. Division impedes the process, growth, and expansion of the Kingdom of God. throughout my Christian walk, I have bee exposed to various doctrinal beliefs, by sitting under the teachings of Apostolic, Baptist, Four Square, Church of God in Christ, and nondenominational congregational church teachings. And, I perceive that many of our doctrinal disagreements are due to a lack of understanding, limited communicational skills, inferiority complexes, fear of the unknown, pride, stubbornness, and resentfulness. I realize that in many ways we're saying the same things, but using different descriptive words in saying them. But if Christians sincerely listened to each other, they may find that there are some similar points of expression. This is just like a couple that has a disagreement, then later for the sake of the relationship, they then decide to consider each other's opinion, for the benefit of continued harmony. Instead, Christians should stop the hostility and slanderous statements, and instead, begin to grow together in love, godly wisdom, and the spirit of peace. Don't allow vain imaginations to hinder spiritual camaraderie and growth, by casting mean - spirited aspersions upon each other. Believers ought to be aware that there's a shared common enemy among all of us. Moreover, that this division is a result of spiritual division within high places. However, this enemy, the devil can transform himself to appear to be an angel of light, and comes in the form of religion that's contrary to the truth. The adversary's goal is to divide and conquer from the inside out. This attack is specifically directed at congregational believers in the faith. However, scripture warns believers not to be tossed to and fro by every wind of doctrine. Because not every doctrinal belief is to be embraced by the Christian faith. Some doctrines are heresies, and their purpose is to deceive many towards opposing the Gospel truth. That is not to say that the poor individuals who have fallen into heretic practices or ungodly beliefs, cannot be won over and converted by the Word of God. Aquila and Priscilla shared enlightenment with a Jewish man named Apollos. Even though Apollos was eloquent and mighty within the scriptures, he was not reticent towards receiving correction and spiritual enlightenment. Apollos had been instructed in the way of the Jesus, and spoke boldly in the synagogue, though he knew only the baptism of John. When Aquila and Priscilla heard Apollos preach, they took him to the side and explained the way of God more accurately to him (Acts 18:24-27).

No one Christian denomination has all of the answers, but each has something valuable to contribute to the growth of the body of Christ. Know that while sharing the Gospel truth, all may not receive you openly, and may verily reject what is being taught or preached. In the presence of the Lord there's liberty and godly truth that can make you free. Apostle Peter, through a vision from God, received the revelation of the truth concerning the expansion of the Gospel being preached to include the Gentiles that are in the faith.

Believers who are filled with the Holy Spirit, and operating in God's love, can witness and express Christ to those who are in darkness. Believers can prove what is righteous, by their words, actions, and deeds. Through the power of the Holy Ghost, the believer can beckon for those who may be lost, to instead be drawn to the transforming light of Christ.

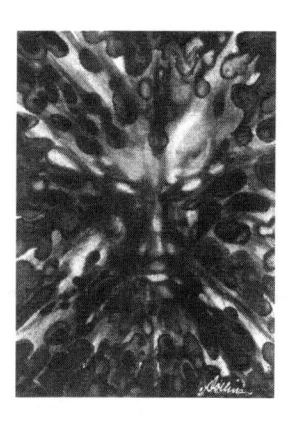

Face of Lies, Deceits, and Darkness

The face of darkness is Satan the devil. And from that face evolves sin, which is void of the presence of God. Indulgence to sin is not a liberality, but instead carries the burden of bondage and sets the stage for a life set on the course of destruction.

The devil IS ALREADY DEFEATED, and doesn't fully understand the plan and purposes of God, because those privileges are beyond the scope of any outside influence that are under Satan's suggestive impact. This denomic adversary, the devil, can never be like God. Even though, he has tried for eons without any success. However, this chief fallen angel called Satan, is the father of all lies, and he is deceitfully crafty at producing counterfeits. These counterfeits are: false angelic beings, antichrists, false deities, fake priests, false prophets, bogus pastors, demonic worship, phony salvation, false doctrines, false believers, familiar spirits, false security, deceitful gifts, liars, perversions, diverted truth and idol worship. Through his spirit of deception, many are deceived to follow after the devices of this evil being. Because of the spirit of darkness, many are lead astray unto total damnation and eternal destruction. All of this is in conjunction with a worldly system that's based upon lies. This adversary is jealous and angry, and is relentlessly driven towards killing, stealing, and destroying what belongs to God. Above everything else, the devil desires to be worshipped and to possess those who will trust and obey him. And he has always coveted to set himself up to be God, with no avail.

God's people have a common spiritual foe, that is, the "devil." This foe rages a constant battle to infiltrate and control our thought processes.

The objective of the enemy is to strive to enable a carnal or ungodly nature that aggressively tries to overpower the spiritually renewed self and godly nature. The solution for this battle is to remain under the shadow of the Almighty God; through His Word, while having the mind renewed daily and transformed into the Mind of Christ. You will then move in power and develop a heightened level of maturity, which enables you to resist the enemy until he runs away or takes flight! This battle is not against flesh and blood, but against spiritual wickedness in high places and pulling down of strongholds, and every high thought that exalts itself against the knowledge of God. A believer is not only just a conqueror, but a victor as well!

"No weapon formed against us shall prosper."

Then he called his twelve disciples together, and gave them power and authority over all devils, and to cure diseases. (Luke 9:1)

Our God Is a God of Love and Order

Born again believers live in a duel social system during this present time of dispensation. There's a worldly system, which is the backdrop of an unholy belief system. The very opposite would be a system of Kingdom Dynamics that is governed by godly remaining principles of love and order. Christians are to operate under a heavenly mind-set and Kingdom Order. An order that operates under the guidelines and dictates of the Lord. No longer should the believer consider being a part of this present worldly system, but should realize that there's an inherited heavenly system that's set apart, even though we still remain within the world. Elevate to having a mind-set that the source of all that's deemed as desirable and needful is in God the Father. And by the Lord's divine orchestration we live and have our being. You must also understand that each person has a specific plan and purpose, for their individual and collective lives. To live in this purpose we must act as an intricate piece of the Body. Also, we are to carry out a genetic characteristic trait of expression of our Heavenly Creator. All of His traits can be directly attributed to the ultimate summation of the Power of His Love. But this loving God also hates sin. Choosing to act out in sin demonstrates unbelief toward what God says is true, righteous, and holy. Rebellion opposes what God says, and consciously responds or acts contrary to His plans and purposes. Unbelief and rebellion strive to bring disorder into the Body. Faith opposes unbelief, and therefore brings the Body in alignment to the order of God the Father.

Power without discipline, leads to destruction because of its chaotic and rebellious forces. But even ungodly power has its limits to what it can or cannot do, but is subject to God's permission and allocation. However, the Holy Spirit enables self control and discipline to occur within the life of each believer to live in victory. Therefore, by being filled with the Holy Ghost there's power to resist the devil, bind the enemy, and render any evil devices ineffective. The Power of the Holy Spirit is always righteous and true, and is never at odds with the Will of the Heavenly Father, but only serves to operate in agreement with the Father and Son. The indwelling Power of the Holy Ghost will only lead you to do what is righteous and true. This ultimate Power will enable you to reach beyond any normal limitations. The Pentecost has initiated the ground work for building the Kingdom. Jesus Christ is the chief corner stone of our foundational principles and doctrines that were laid, and now stands. According to Isaiah 45:1, God's anointed are given believers the authority to subdue nations by His power.

*And God blessed them, and God said unto them, Be fruitful, and multiply, and replenish the earth, and **subdue** it: and have dominion over the fish of the sea, and over the fowl of the air, and over every living thing that moveth upon the earth.* (Gen. 1:28)

God's heavenly angels hearken to His every word, and they hasten to perform it. Believers can also call upon His angels, by speaking the words of God from a pure heart.

Spiritual and Natural

Our natural and earthly genetics were tampered with sin by way of the First Adam, which then caused the entire human race to inherit a generational curse. But God, through his Saving Grace and Mercy, along with the shedded blood of his Only Begotten Son, causes those who believe to receive salvation. Christ's crucifixion allows the believer to inherit His blood covenant, and to receive a new transfusion of blood purification and life resurrection into new godly genetic traits. By Grace, Believers are then made children of God, whereby they are able to possess the new nature that comes with being a new creature in Christ. Having been equipped with the same *Spiritual DNA* as Jesus Christ. Born again believers carry the genetic coding of our Creator, having characteristic traits that are similar to those of the Heavenly Father. This enables godly characteristics to rise within the believer, and encourages the believer to follow after God with fervor and conviction. This empowers children of God not to govern themselves after vain imaginations, but to instead acknowledge the Lord in all ways by honoring His dictates and precepts. As children of God, the mind is transformed so that it thinks like unto the Heavenly Father. The born again believer takes on a new nature to receive its instructions to operate in alignment with the Mind of God. The believer then processes the godly characteristics to serve, give, sacrifice, and love as one people in Christ Jesus. We as corporate-spirit-filled believers, are divinely infused together into One Body. Born again believers are re-configured into the Body of Christ that anxiously awaits us as becoming the Bride of Christ.

Our universal and supreme **_Spiritual Headship,_** who primarily holds all authority, is God Himself. The Spiritual Headship holds all spiritual, physical, and psychological authority over the mind, body, soul, and spirit. Jesus Christ is that Headship, and His Glory is the Crown that sits above the Body of Christ. The Spiritual Head is the authority of God, which governs the entire Body of Christ. The Spiritual Brain is the Word of God. The Mind of Christ is the illuminated, unadulterated, and uncompromised spiritual consciousness toward the wisdom and council of God.

The **_natural eye_** is an organ of vision or of light sensitivity located in bony sockets of the skull. The eyes function together or independently, each having a lens that is capable of focusing incident light on an internal photosensitive retina from which nerve impulses are sent to the brain. The brain of course, serves as the vertebrae organ of vision. This is the external and visible portion of this organ, which along with its associated structures within the eyelids, eyelashes, and eyebrows. The **_eyelids_** are two folds of skin that can be moved to cover or open the eyes. Even when the eyes are open they still remain protected by a thin layer of membrane that covers each eye. Lashes act as screens. The **_eyebrows_** are an arch of hair above each eye. Dense growth of **_hairs_** covering the body or parts of it (as on the human head) helps prevent heat loss.

Spiritual and Natural

The physicality of the **_Spiritual Head_** is equipped with eyes to see with. These **_Spiritual Eyes_** give us the ability to make wise, intellectual, or aesthetic judgments, and to remain focused on the vision of God. In other words, spiritual eyes enable believers to utilize spiritual sight in order to decipher spiritual and natural happenings, conditions, situations, circumstances, and even their destiny within God. We are able to see Him for who He is. We are able to visualize what is true, pure, and holy. The awareness also has a way of allowing us to discern whether something comes from a godly mind-set or not. Having spiritual sight gives us the ability to see into the matter of spiritual things, and to conceptualize God's laws, precepts, and ordinates. This is necessary in order for the believer to receive spiritual direction and to operate by faith in Jesus Christ. Having spiritual eyes gives us the ability to have insight, hindsight, and foresight into the future, in addition to possessing knowledge concerning eternal matters. Otherwise, being spiritually blinded is the absence of faith, making it impossible to please the One true God. Therefore, it is also impossible to be fitly joined together in fellowship with other members in the body of Christ. Lack of faith in God means, that one is spiritually blind and void of fellowship with God the Father. Most importantly, without the God kind of faith, one is alienated from the Source of eternal truth.

When it comes to the ***Spiritual Anatomy***, nothing is insignificant. Even the ***Spiritual Eyebrows*** help to retain what is perceived through the word and vision of God. Even eyelashes aide in filtering out invasive visual seductions from evil influences, by opposing ungodly influences from entering through the eye gates. The ***Spiritual Eyelids*** protect creative imagination and garner what is perceived through visionary effects, patterns, and boundaries of the word of God.

Spiritual and Natural

Our ***natural ears*** are the vertebrate organ for hearing, and they share responsibility for maintaining equilibrium, along with sensing sound. The ear is divided into three parts: the external ear, the middle ear, and the inner ear. *The **external ear*** is the outer portion of the ear, including the auricle, and the passage leading to the eardrum. The ***eardrum*** is the thin, semitransparent, oval-shaped membrane in the ear that vibrates to sound, and separates the middle ear from the external ear. The **middle ear** is the space between the eardrum and the inner ear, which contains the three auditory ossicles, which convey vibrations through the oval window onwards to the cochlea. The ear is an organ having nerve endings, which respond to stimulation. The ***inner ear*** is located within the temporal bone. It is involved in both hearing and balance and includes the semicircular canals, vestibule, and cochlea. The ear is one of the vertebrate organs that are externally visible. ***Hearing*** is to have a perceived auditory sense of sound.

The **Corporate Church Body's *Spiritual Ears*** are necessary for hearing and recognizing the voice of God. They are also necessary for godly receptiveness and awareness to what is holy. The spiritual ear has a refined sense of discernment and receptiveness to what is heard. It even has a vestibule that's similar to the natural ear. Christian believers possessing character traits that are relative to the spiritual ear serves with qualities and activities of a priestly position, through sanctification, worship, and prayer as it harkens to hear the Voice of God. Spiritual leaders, such as bishops, pastors, teachers, prophets, and evangelists must have spiritual hearing in order to hear God's voice for communication, instructions, inspiration and fellowship. It would be impossible to lead others in a godly path without being attuned to spiritual hearing. Actually, all born again believers should possess the trait of spiritual hearing in order to listen to God's voice, and then understand what is being spoken. Its ***Spiritual External Ear*** is the outer threshold and preparation leading to sensitivity and receptiveness of the spoken Word of God. The ***Spiritual Eardrum*** is the part that is keenly receptive to the voice of God, so that the word uttered may be sown into the hearts and minds of every believer. The ***Spiritual Middle Ear*** is the seat of receptiveness, reflectivity, and discernment. This is also a place of revelatory expression. A ***Spiritual Inner Ear vestibule*** is the central cavity leading to another cavity. The vestibule also has similarities to a church vestibule, which is a holding place or passageway where sanctification takes place before entering into other cavities or passages. This is an anointed threshing place, where all communicated messages that are received must first be deemed as being truly sent from God. There is a process of proving what is heard to be true and holy before there is access into its sanctified place. The process of sanctification will shut out all interference of that which opposes the distinct Word and Will of God. The Lord tells believers that our spiritual ears will ***hear*** and recognize His voice, and will not receive nor follow after a counterfeit as the true voice of God. From an individual standpoint as a believer, having spiritual ears will enable a believer to follow after Christ by adhering to the voice and dictates of God in order to walk circumspectly, according to His ordinates. Without spiritual ears, one can neither receive God's instructions nor obey Him. Furthermore, spiritual deafness occurs due to the absence of faith. Believers are supposed to operate by faith. Because, faith comes by hearing the Word of God.

Spiritual and Natural

The ***nose*** is the part of the human face or the forward part of the head of other vertebrates that contain the nostrils and organs of smell, while also forming the beginning of the respiratory tract. The anatomy of a tract is a system of organs and tissues that work together to perform a specialized function: the alimentary tract. A tract is also a bundle of nerve fibers having a common origin and terminus, in addition to functioning as a system of body parts that work together to serve some particular purpose. A system is a group of physiologically or anatomically related organs or parts. The respiratory tract is the passageway through which air enters and leaves the body. The nose, throat, and trachea are parts of the upper respiratory tract. A nose's natural purpose would be to detect and discover through the sense of smell, and by means of the olfactory nerves. The sense of smell is activated by olfactory receptors in the nose, which are stimulated by particular chemicals in gaseous form.

The **Celestial Nose** has spiritual sensory receptors and fibers, which are similar to the olfactory receptors and fibers in the human nose. Its purpose is to have a heightened sense of awareness concerning present surroundings, and to detect when something is favorable, or not quite right. Moreover, the Celestial Nose senses when the enemy is near. This is a system for discernment. This member's main objective is to receive fresh inspiration from the Word of God, and to transmit to other parts of the body.

Primarily, the qualities and characteristics of spiritual sensory receptors and fibers are that they must remain sensitive to the nature of God, while still remaining resilient to outward distractions. Even during times of testing, these receptors still remain focused towards the mission of transmitting and declaring the message from God, declaring the power of the Holy Spirit through prophetic utterances.

Spiritual and Natural

The anatomy of the **human mouth** is the cavity lying at the upper end of the alimentary canal, bounded on the outside by the lips, and inside by the oropharynx. It contains higher vertebrates within the tongue, gums, and teeth. This cavity is regarded as the source of sound and speech. The part of the mouth that's visible on the face are the lips. The lips are either of two fleshy folds that surround the opening of the mouth.

That which is the **Spiritual Mouth** reflects and represents the spoken Word of God. Even the lips are formed to embrace the very essence of God's breath. The spiritual tongue tastes the goodness of the word, and echoes or repeats what the Word of God has already spoken into existence through the eons of time, by way of the Holy Ghost.

A **human tooth** is one of a set of hard bonelike structures rooted in sockets within the jaws, used for biting and chewing, or for attack and defense. Teeth are more than just one tooth in a set arrangement along the gum line. The jaw is either of two bony or cartilaginous structures that in most vertebrates form the framework of the mouth, while also holding the teeth. The chin serves as the central forward portion of the jaw. The gum is a firm, connective tissue covered by mucous membrane that enfolds the arches of the jaw, and confines the bases of the teeth. And the tongue is a fleshy, muscular organ that is attached to the floor of the mouth. This is the principal organ of taste, and it also aids in chewing, swallowing and articulating speech.

The Spiritual Anatomy has **Spiritual Teeth** that are set up within Spiritual Gums that are hinged upon a Spiritual Jawbone. Its spiritual teeth are held in strong alignment to perceptively disseminate the word of God. And to break down spiritual substance and aid ingestion to support its digestive process. Spiritual Teeth chew upon the Bread of Life in order to feed off it for the life-giving force that is within it. From this, the operational discernment gifting of wisdom brings knowledge and understanding, that bare witness to revelation of godly truth. These Spiritual Teeth are characterized like unto watchmen, who watch, or like council who advise and act as seers to the kingdom-building work. Spiritual Teeth also act as spiritual gates that guard against that which attempts to enter, in addition to eliminating that which goes out to defile the temple of God. Those bearing this trait are teachers, prophets, and councilors who bring clarity to God's word and voice, by the enabling wisdom and knowledge of the Holy Spirit. This work leads and guides unto all truth. Each believer in the Body of Christ must develop and mature in the Lord to grow spiritual teeth, so that the Spiritual Meat can be received from God's word. This enables proper digestion, and ensures that the believer has a healthy and heavier appetite for the Word of truth. The believer can now be moved from the "milk," which is sufficient for babes, and onward towards growing in spiritual maturity. Therefore, the believer is then enabled to partake of and understand the deeper things of God.

Spiritual and Natural

The **Spiritual Gums** support what is being received into the mouth. The jaw then tenaciously holds on to the very meat of the word of God without wavering, and there is no partiality as to how it's received. Those who have these special characteristics of the gums, are the truly ordained overseers and elders of the church. Moreover, these members tend to be more spiritually mature and refined, in addition to being more willing to help move forward the proclamation of the Gospel being preached, towards instructing the Body. More importantly, they declare the gospel outwardly to the world.

The spirituality of the *Celestial Tongue* is a very strong member of the body, possessing the ability to change its environment. It must remain rooted to the sure foundation by adhering to God's foundational truths to draw from. This charters the direction or the way in which the other members will follow. That is why this member of the Body must remain sanctified and holy through consecration and by prayer, which ensures constant adherence to the Will of God. This part of the body must articulate and repeat the words of God in purity and truth. The holy tongue has the power to decree a thing according to the Will of God, through the inner-working of the Holy Spirit.

Characteristics of the *Spiritual Jaw* are typical of those who are steadfast and unmovable in the faith, while also abounding in the faith that actively aligns to prove what's spiritually upright to be true. This member will resist wavering and has the gift of intercession and discernment. The spiritual lips work in a supportive fashion along with the celestial tongue to help formulate with clarity to utter the Word of God.

How Does This Anatomy Relate to Us as Corporate Believers in Christ?

Naturally, the skin acts as a protective layer that helps to guard the body against foreign penetration. Spiritually speaking, if you are a part of the corporate body of believers, then you're obligated to care for the other through prayer, sharing, giving, and loving.

The holy covering of each member of the Body of Christ is the Blood of Jesus by the Word of God, that invariably works to guard and sanctify the body, inner spirit, and soul for each born again believer. There are many levels and chambers of sanctification that inevitably lead to God within the Body of Christ. But there's no name other than Jesus Christ, whereby we can be saved. The sanctifier is the Holy Spirit of God. The Holy Ghost is housed in the temple, which is the heart of the body. The very heart of God is Jesus Christ. The Bible proclaim that "where your treasure is, so is your heart also." The God Almighty "El-Shaddai," gave us what was most near and dear to Him. He gave us His Heart.

The **Body of Christ**, also called the **Church of God**, should stand as a reflective image of God's Love that was manifested through Jesus Christ. Thus, Jesus is the Word of God, and is activated through our Lord and Savior by way of translation through the Holy Spirit. What was already settled in heaven is now being established here on earth! The Body of

Believers are to be in direct agreement with what was spoken and yet already is, by the Word of God. Believers are to adhere, trust, and obey the Word to steadliy remain within the Will of God. By way of words, actions, and deeds we are to represent what has already taken place in heaven. The Lord tells us to let His will be done on earth as it is in heaven (Luke 11:12).

Spiritual and Natural

In our observation of how the anatomy of the human body moves and operates, members can also sense the relative operation of commands and statutes of God for the corporate Body of believers. Here are other descriptions of parts and their interactivity, as well as their relational values with other counterparts within the physical human and spiritual anatomy of both bodies. To save space and time, for the remainder of the other anatomical parts, I will refer to the **human anatomy** as "*HA*" and the **spiritual anatomy** as "*SA*."

Human anatomy descriptions are taken from "TheFreeDictionary.com" under the GNU Free Documentation License.

The Spiritual Anatomy is referred to as the Body of Christ, Body of believers, Bride of Christ, The Church, and Corporate Body of believers. The references and descriptions are taken from Bible sources which include the: (King James version, New International Readers version, New Living Translation, and American Standard Translation.)

Spiritual and Natural

HA

The ***Throat*** is the portion of the digestive tract that lies between the rear of the mouth and the esophagus, and includes the fauces, pharynx, and the neck. It is also the passageway to the stomach and lungs. It is in the front part of the neck below the chin, and above the collarbone.

SA

Throat: The Spiritual Throat welcomes the Word of God through worship, praises, and consecrated prayer in order to receive what is the word of faith. This member is a component for receiving the Will of God, and a supporter through the various means of declaring what God says. Typical vassals who possess the Spiritual Throat trait would be: pastors, teachers, prophets, evangelists, intercessors, and other consecrated believers.

HA

Tonsils are two small oral masses of lymphoid tissue embedded in the lateral walls of the opening between the mouth and the pharynx, which may perform unspecified functions, but they are believed to help protect the body from respiratory infections.

SA

Tonsils: Spiritually speaking, the characteristics may be relative to those that are positioned in part as ***Celestial Tonsils***, although they may appear as insignificant and considered to be dispensable. In truth, they are no doubt very valuable to the harmonious balance of the Body. This is characteristic of those whose work is perceived to be more on a passive level, and not well defined, but yet still very necessary.

HA

The Neck is the part of the body joining the head to the shoulders or trunk. **The Thymus Gland** is a ductless glandular organ at the base of the neck that produces lymphocytes and aids in providing immunity. It atrophies with age. The **Arterial carotid** is either of two major arteries of the neck and head; it branches from the **aorta**. The **aorta** is the main trunk of the systemic arteries, carrying blood from the left side of the heart to the arteries of all limbs and organs except the lungs. The ***Jugular*** is a vein in the neck that returns blood from the head. **Trachea**, also known as a windpipe, is a membranous tube with cartilaginous rings that conveys inhaled air from the larynx unto the bronchi. The **Neck bone** is one of the seven cervical vertebrae (located in the neck region) in the human spine.

SA

Neck: In the Body of Christ is the ***Spiritual Neck*** that turns its face toward God. Its aim is to seek God through worship in submission, and obedience. The neck also bows to humble itself before God, while worshiping in spirit and in truth. Those who carry these traits are Orthodox Jews and Spirit-Filled Christians that are operating in positions such as: pastors, godly priests, missionaries, and intercessors.

Spiritual and Natural

SA

The Spiritual Thymus gland of the Body of Christ is a spiritual immune system that does spiritual warfare through the power of intercessory prayer. These would be typical traits of this gland's operational gifting within the Spiritual Body. Characteristic of those who oppose any intrusion from adversarial attacks that would enter the Body to fester forms of alienation and spiritual sickness.

SA

Spiritual Arterial Carotid: This member is typical of those who remain connected to the True Vine of God, bearing much fruit of its kind to harvest blood-washed disciples. There are no hybrids or counterfeits here, because these vassals stick close to the apostolic doctrinal teachings.

SA

Spiritual Aorta: Believers having the typical traits of this member are also makers of other disciples, by the various gifts of evangelism. They preach and teach the Acts 2:38 message of salvation continually, almost like a broken record. They cry out to all who will hear. By the anointed preaching of the Gospel they exhibit the power of God through the excising and demonstration of spiritual gifts to win souls.

SA

Spiritual Jugular: This member allows the Blood to flow through the Body from HEAD to Toe. This trait is similar to those who shun the very thought of grieving the Holy Spirit. The Spiritual Jugular has character traits that are typical of those who are always looking for the presence of God in all things. This type of member diligently seeks to acknowledge the Lord with earnest desire for the revelatory experience of God. These vassals also encourage others to be open and sensitive to the manifestations of the move of God. They thrive to remain connected to the Heart of God, and are inspirationally compelled by the thrust of His Heart beat. And remain closely sensitive to the pulse of His Hand.

SA

Spiritual Trachea: This member's relative function is to ensure that the fresh **breath of the Spirit of God** *is* inhaled to receive the **Life given force through His Word.** Vassals carrying this trait are those who possess talents such as: inspiration, motivation, and charisma. Their gifting could be in the area of word of knowledge, revelatory knowledge, prophesy and extraordinary faith.

SA

Spiritual Neck bone: This position is relative to those who labor in assistance to support ministerial headship.

HA

Digestive tract is a tubular passage of mucous membrane and muscle extending about 8.3 meters from the mouth to the anus. It assists in digestion and waste elimination.

SA

Spiritual Digestive Tract: within the spiritual body this line of defense must detect what is to be received into the body as God's nurturing and life supporting properties. While rejecting or eliminating whatever is deemed to be detrimental for the Body's consumption.

Spiritual and Natural

HA

The Stomach is an enlarged saclike portion of the digestive track between the esophagus and small intestine, lying just beneath the diaphragm.

SA

Spiritual Stomach: This is where the center of desire or passion lies, and the center of where worship and prayer derive. It is also where the Body receives the Word to digest it for all of its nutritional values through spiritual interception. Then spiritual interpretation takes place to process its various values for distribution of its spiritual nutrients throughout the Spiritual Anatomy. It's relative to those having characteristics that pave and assist the way for strength, energy, and vitality throughout the Body. This is done through the Power of the Holy Ghost.

God's pure unadulterated Word is completely good, and comes with complete nutritional values that symbolizes every facet of wholeness for us. The Holy Ghost produces spiritual enzymes to help process the spiritual bread of life, so that the Body can properly digest it through proper interpretation for strength and spiritual health. However, it's apparent that the devil does exist and devises to infiltrate the church body through the spreading of deception. Then the devil's deceptions must be filtered by the Word of God, and brought through the process of elimination to discard anything that's tainted by outward negative elements, which some weakened vassals may allow to come in. It is also the place where nurturance and godly qualities are drawn from the word of God, in order to translate it to the various parts of the body by the purification and efficacious process of the blood of Jesus. Within the stomach lies the infiltration of qualitative, qualitative and quantitative values, along with the traits of God. These traits are carried through the spiritual blood stream, and then is circulated throughout the entire Spiritual Body.

Within the Anatomy of the Spiritual Body Is a Soul

This is where the center of deep-seated passions rest within God. This works to bring elation and peace to our spiritual mind and body. Within the Soul lies the innermost place where our shared intimacies toward God through Christ exists, along with our love of His Holy Word. Every living person has an eternal soul. A soul is the immaterial part of the self that expresses righteous conduct or evil desires. The spirit of a person can be defined as their nature, belief system, or controlling influences. The spirit is also the center of our emotions, the source of passions, cause of will, and is the subject of divine influence. The spirit is also the life force, nature, and character of who and what a person is made of.

A redeemed soul exercises a life of righteous living and conduct, and therefore bypasses evil. The soul is comprised of intellect, emotions, and volition. A soul also carries traits and information concerning the past, present, and future spiritual life. It will retain information relating to a person's physiological, emotional, and intellectual state, while also establishing the nature that will determine an individual's eternal chartered destination. This will subsequently determine where the soul will spend eternity; either in heaven or in hell. The soul is capable of sinning, and is thus deemed responsible for carrying out the volition to sin. It is a valuable asset, and therefore everyone must be careful not to live a life of sin that would later bring about judgment for the deeds of the flesh. Only God can see into the soul; the spiritual life that has been placed there, and the course of life that is to be carried out. This holds the true sense of who the individual really is in terms of intellect, personality, and desire. The original state of the soul before taking on the sinful nature of Adam (after the fall of man), was one that contained the spiritual genetic coding or godly characteristics of the Heavenly Father. The spirit of mankind first originated from God. He breathed out from Himself, the life giving force to generate a living soul. Then placed the soul into the physical body of Adam. And thus began the human race. Prior to entering the physical state or earthly realm, our spirits were preexistent in heaven with God, before inhabiting earthly body suits that were later formed in our mother's wombs by God. However, humankind's sinful nature began doing the fall of Adam's original sin, but thank God for the second Adam as Jesus Christ. Through His work of redemption, He was able to reconcile and restore us back to the original nature of God, by way of the Holy Spirit. He then offered His provisions for salvation by the sacrificial death of Christ. And most importantly, the blood that was shed for the remission of every believer's sins. On the other hand, the unsaved souls are spiritually disconnected from God, and thus remains carnal or fleshly in their thinking, inclinations, desires, or expressions. Therefore, leaving the unsaved soul in a state of hopelessness and destruction. The Spiritual Body also has a Spiritual Soul that's holy and dwells within a Soulful realm.

The Spiritual Soul is within the Spirit of Christ, that is, the personification of who Christ truly is. Its Soulful realm is where the Spiritual Soul dwells or inhabits. The Holy Spirit is with God, and He inhibits the redeemed souls of every believer. The Spiritual Body incorporates the Spiritual Heart, Soul, and Mind of Christ. If you are born again into the Spiritual Body of Christ, you then have the ability to boldly approach God from a position of authority. You must become one with Christ in order to live through Christ. Furthermore, you are to love God righteously, by loving Him with all of your mind, body, and soul in the power of the Holy Ghost. Without this synergistic connection, there can be no true fellowship or relationship with God. Because our lives are hidden in Christ, it is no longer I that live, but Christ that lives within each of us, and to whom our souls have been redeemed. The redeemed soul is rescued and renewed in the Soul of Christ. A redeemed soul is wrapped up in the Soul of Christ, and begins to take on the personality and expressions of Christ. Our redemptive soul is full of hope and glory through Christ. The traits of those who dwell in the Soulful realm of the Body of Christ, are those who are spiritually vibrant and passionate leaders. This list would include sanctified pastors, teachers, praise leaders, and true worshippers.

Spiritual and Natural

HA

Bones are a dense, semirigid, porous, and calcified connective tissue that forms the major portion of the skeleton of most vertebrates. They consists of a dense organic matrix and an inorganic mineral component. There are more than two hundred different bones in the human body.

SA

Celestial Bones: They structurally support the Spiritual Anatomy, enabling it to stand erect and to walk. This would be typical traits of those who support the growth and health of a ministry. Many times they operate in the gifting of intercession, or within the area of spiritual counseling (helping). This can be extended to include offering of their gifts, talents, or financial support.

HA

Ribs: Are a series of curved bones that are articulated with the vertebrae, and occur in pairs—twelve in humans—on each side of the vertebrate body. Certain pairs connect with the sternum to form the thoracic wall.

SA

Ribs: Relative traits have the function of protecting the spiritual respiratory system, to guard its dominion, and to expand or broaden its territory of the Kingdom. Other relative characteristic traits are found in the twelve tribes of Judah, twelve foundations of the City of God, twelve gates, twelve Apostles of the Lamb, Apostolic order within the Kingdom. For the Body of Christ, the spiritual ribs represent the **Kingdom Order of God.** Spiritual Ribs are positioned as members of authority to uphold godly truth and standards, as measured by the **Word of God.**

The Son of Man Is Given Dominion

I saw in the night visions,
and behold, with the clouds of heaven
there came one like a son of man,
and he came to the Ancient of Days
and was presented before him.
And to him was given dominion
and glory and a kingdom,
that all peoples, nations, and languages
should serve him;
his dominion is an everlasting dominion,
which shall not pass away,
and his kingdom one
that shall not be destroyed. (Dan. 7:13, 14)

New Jerusalem, the Holy City of God (Rev. 21:2, 10-16, 24-27)

And I John saw the holy city, new Jerusalem, coming down from God out of heaven, prepared as a bride adorned for her husband. (Rev, 21: 2)

And he carried me away in the spirit to a great and high mountain, and showed me that great city, the holy Jerusalem, descending out of heaven from God, Having the glory of God: and her light was like unto a stone most precious, even like a jasper stone, clear as crystal; And had a wall great and high, and had twelve gates, and at the gates twelve angels, and names written thereon, which are the names of the twelve tribes of the children of Israel: On the east three gates; on the north three gates; on the south three gates; and on the west three gates. And the wall of the city had twelve foundations, and in them the names of the twelve apostles of the Lamb. And he that talked with me had a golden reed to measure the city, and the gates thereof, and the wall thereof. And the city lieth foursquare, and the length is as large as the breadth: and he measured the city with the reed, twelve thousand furlongs. The length and the breadth and the height of it are equal. (Rev, 21: 10-16)

And the nations of them which are saved shall walk in the light of it: and the kings of the earth do bring their glory and honour into it. And the gates of it shall not be shut at all by day: for there shall be no night there. And they shall bring the glory and honour of the nations into it. And there shall in no wise enter into it anything that defileth, neither whatsoever worketh abomination, or maketh a lie: but they which are written in the Lamb's book of life. (Rev. 21:24-27)

Spiritual and Natural

HA

Joint: The point articulation between two or more bones, specifically, those connections that allow motion.

SA

Spiritual Joint: This is a member that joins together with another in the manner by which the Bible tells us to be knitted together in love. Love is the cohesion that binds two or more together to work as one. Joined members of the Body are to be like-minded, and work in cohesion to perform the same objective collectively.

HA

Muscles: are tissues composed of fibers, which are capable of contracting to effect bodily movement. A muscle is a contractile organ consisting of a special bundle of muscle tissue, which moves a particular bone, part, or substance for the body.

SA

Spiritual Muscles: typically these are motivators who strengthen core values, maintain moral ethical behavior, and establish spiritual integrity. Additionally, they promote commitment, movement, and activity within the Body. Pastors should seek to possess this trait, which comes by the anointed power of influence. By operating and holding the office to preach and teach, they echo God's words that produce the faith that strengthens the Body of believers. By their God-given strength, those that hear the Voice of God through these vassals are themselves made stronger.

These muscles receive their vitality and stamina from the nutritional value through the word of God. This builds them up to be strong by praying in the Holy Ghost, whereby they are able to live a life of obedience to the Lord and through retroactive faith in God's truth.

HA

Sinew or Tendon: are cords or bands of inelastic tissue connecting a muscle with its bony attachment.

SA

Spiritual Sinew or Tendon: would correspond to a layperson that works to support and promote evangelistic awareness and missionary efforts for community or world outreach.

HA

Ligament: is a sheet or band of tough fibrous tissue that connects bones, cartilages, supporting muscles, or organs.

SA

Spiritual Ligament: this member is composed of those who work diligently to support other ministries and projects. Their goal and mission is to assist and build upon the sure foundation.

HA

Diaphragm, Midriff (anatomy): is a muscular partition separating the abdominal and thoracic cavities; it functions in respiration.

SA

Diaphragm, Midriff: This section of the spiritual anatomy could very well be vassals of the fivefold apostolic ministerial gifting. Its powerful manifestation works to deliver the oppressed, and to provide the encouragement needed to uplift the downtrodden. The functioning power of this member can also serve the sick, and preach the message of salvation.

HA

Tissue: is a part of an organism consisting of an aggregate of cells having a similar structure and function. The Cell is the basic structural and functional unit of all organisms. Cells may form colonies or tissues.

SA

Celestial Tissue: Within the anatomy of the Spiritual Body, is a culmination of similar ministerial colonies that are structurally and functionally similar. They are joined together to serve a common purpose in works and goals. By virtue of the intensity related to their various unified works of commitment, the spiritual muscles are then developed. These traits would be characterized by the various church groups, communities, and denominations. Now bear in mind that unless the various entities come together in love, commitment, and unity, they may soon break away. Alienation of any structural unit forms such as cell tissue, is commonly referred to as cancer. The Body of Christ cannot tolerate the continuous festering of this sort of spiritual cancer, or any other diseased element to enter or exist within. This must be swiftly dealt with through repentance and receptiveness to the mind-set of Christ, and the voice of God. This will promote healing and restoration to the estranged individual to become one with the Global Church body.

HA

Fat: is a kind of tissue containing stored fat that serves as a source of energy. Adipose tissue also cushions and insulates vital organs. Fatty tissue also protects the body from severe cold.

SA

Spiritual Fat: this type of member acts as a spiritual insulation, energizer and protective spiritual buffering agent through the Power of the Holy Ghost, which is also the source of power and regeneration for the Spiritual Anatomy of Christ.

HA

Blood: is a fluid consisting of plasma, blood cells, and platelets that are circulated by the heart through the vertebrate vascular system. This fluid carries oxygen, nutrients, and waste material away from all body tissues.

SA

Blood: In the Spiritual Body, the **Blood** is the *same Blood that perpetuates from the sacrificial Lamb of God, being none other than Jesus the Christ. The Blood is the Life of the Body. Without the shedding of blood, there is no remission of sins. The Blood of Jesus permeates* the Body of believers. *His Blood protects, purifies, and sustains life.*

HA

Blood Vessel: This is a vessel in which blood circulates; the vessel carries blood from the heart to the body.

SA

Spiritual Blood Vassals: This is a relative characteristic of the vassals that continuously preach and teach the unadulterated Gospel of Jesus Christ. They serve as connecting agents of salvation towards the redemptive work of Jesus Christ.

HA

Artery: Any of the muscular elastic tubes that form a branching system, and carries blood away from the heart to the cells, tissues, and organs of the body.

SA

Spiritual Body Arteries: They are attached to the Spiritual Root of Jesus Christ. These Spiritual Arteries are connected to the *divine source of Holiness, Truth, and Power*. Through the *Living Word, speaks forth the Light of the world. This enables the Church Body to remain divinely connected to its source Jesus Christ.*

HA

Cell: is the basic structural and functional unit of all organisms and cells, and may exist as independent units, or may form colonies or tissues. It is the smallest structural unit of an organism that is capable of independent functioning, and consists of one or more nuclei, cytoplasm, and various organelles, all surrounded by a semipermeable cell membrane.

SA

Spiritual Cell Unit: is a church fellowship group of any size.

HA

Nucleus (nuclei) *n. pl.* or (nucleuses)

This is a central or essential part, around which other parts are gathered or grouped.

The ***nucleus:*** is a large, membrane-bound, usually spherical protoplasmic structure within a living cell, that contains the cell's hereditary material, and controls its metabolism, growth, and reproduction.

SA

Church Nucleus: this serves as a core connective unit of spiritual headship. This is an operating pastoral position within a church unit. From the spiritual nucleus abides the hereditary material that controls spiritual metabolism, growth, and reproduction of the church unit. This leadership position sets the tone and belief system of its cell group of believers. This apostolic position of authority provides a protective covering for its immediate members, under its leadership.

HA

Protoplasm: is the living substance of a cell (including cytoplasm and nucleus). It is a complex, semifluid, and translucent substance that contains the living matter of cells. Addtionally, this substance manifests the essential life functions of a cell. Composed of proteins, fats, and other molecules suspended in water, it also includes the nucleus and cytoplasm.

SA

Church Protoplasm: this serves as the spiritual content that constitutes the nature of its reproducing life-giving force that emanates through out a church cell unit. The manifestations of the life of each fellowship group will exemplify and replicate the nature, principles, and workings, whether they be good or bad.

It will produce of its own kind. However, a cancerous or defiant church cell units cannot thrive within a holy environment. Within the Body of Christ, unholy cell units are either purged for purification purposes or broken off and eradicated.

HA

Cytoplasm: is the protoplasm of a cell excluding the nucleus of a cell.

SA

Cytoplasm Church: This is a dead church cell unit; this is not a true Christian fellowship group. It is a counterfeit church cell unit. It may resemble the typical church unit, but it lacks the true foundation of what The Church Body is established upon. The counterfeit cell is without the anointing of the Holy Ghost, and most importantly without the Savior Jesus Christ as Lord and God, who is the true Life of The Church.

HA

Cell Membrane: is the semipermeable membrane that encloses the cytoplasm of a cell. A thin membrane around the cytoplasm of a cell; it controls the passage of substances in and out of the cell.

SA

Church Cell Membrane: represents the anointed and appointed apostolic authority of God. Through an anointed individual, such as an ordained pastor that covers, protects, and defines a church cell ministry and its members.

HA

Nervous System: is the sensory and control apparatus consisting of a network of nerve cells. A nerve cell or neuron is a cell that is specialized to conduct nerve impulses.

SA

Celestial Nervous System: is the spiritual network or pathway of communication from the Primary Spiritual Head to the global spiritual

Body of believers in Christ. The Body can also communicate via the Holy Ghost to other parts of the Body. In order to communicate back to the Main True Source, which is God Himself. The believer must be connected to the Heavenly Father through the Son, Jesus Christ. God has fashioned the Body to be interdependent upon each other to function as a whole *governing* Body. This works to give Him glory and honor.

The word *Govern* means: 1. To make and administer the public policy and affairs of, or to exercise sovereign authority in. 2. To control the speed or magnitude, in order to regulate. 3. To control the actions or behavior of. 4. To keep under control; to restrain. 5. To exercise a deciding or determining of.

HA

Fiber Bundle: These are fibers (especially nerve fibers). The entire nervous system consists of a central, peripheral, autonomic, sympathetic, and parasympathetic nervous system.

SA

Fiber Bundle: This is what connects any given group through a spiritually effective networking system of communication and joint efforts. This process can initiate out of a mission or visionary effort of a person. It must grow spiritually to produce a result of a related work endeavor that unites others together to serve a cause.

HA

Fiber

(a) This is any of the filaments constituting the extracellular matrix of connective tissue. (b) Any of various elongated cells or threadlike structures, especially a muscle fiber or a nerve fiber.

SA

Fiber: This is the beginning element of a growth process for a spiritually connected networking system. This system is intended to generate a unified work effort that has a godly motivation and direction.

HA

Heart: is a chambered, muscular organ that pumps blood received from the veins into the arteries. This helps to maintain the flow of blood through the entire circulatory system.

SA

Heart: Christ is the true heart of the Body of believers, because out of the love of God came the one true sacrifice for our sins. God gave His only begotten Son to take our place as a sin sacrifice in order to reconcile us back to Himself. This was the ultimate love gift to us. When He gave of Himself, He gave His Heart. From out of His Heart flows Life toward all of us. Our hearts as believers must join with His Heart through repentance from sin and faith in Him as Lord and God. We must love the Lord our God with all of our hearts, minds, and souls. How do you love Him with all of our hearts? You must first submit yourself to Him in the totality of all that you are, and pray to become whatever is delightful in His sight . . . You are to give your thoughts and prayers in times of mediation upon His word and allow His word to change the framework of your mind and in how you think, perceive, and conceptualize thought. Then begin to allow all of your actions to be governed by biblical principles of thinking through the Holy Ghost. The word of God generates a transformed mind, and the Holy Spirit enables the spirit or inner self to receive and comprehend eternal truth. God replaces and gives the believer another heart (1 Sam. 10:9). Furthermore, He gives believers a circumcised heart to love Him in return (Deut. 30:6). The upright will incline their hearts to God, having hearts that are wise and understanding. The condition or integrity of individual hearts, souls, and minds will determine whether or not a person is deemed justified to remain in their positions to be within the Body of believers. If not, they will remain outside of the Body as the unbelieving and unsaved. God gave Judah, the seed of Abraham and the Church, a spiritual heart to do His Will. This is in accordance with the word of the Lord (2 Chron. 30:12). God will give believers the heart to do His Will through a unification of the Mind of Christ, and will transform us to one mind. However, for a time being, He will give our kingdom unto the beast until the words of God should be accomplished (Rev. 17:17). The objective will bring us to one Heart and Mind in the Body of Christ. Then the earthly kingdoms, dominion, and the greatness of the kingdoms under the whole heaven shall be given to the people. As they are saints of the Most High. Their kingdom shall be an everlasting kingdom, and all dominions shall serve and obey them.(Dan. 7:27)

HA

Lungs: They are two saclike organs of respiration that occupy the pulmonary cavity of the thorax, and in which aeration of the blood takes place.

SA

Spiritual Lungs: this spiritual position is relative to those being interactive with the God Source, and in turn, is a part of where and how the Word of God is received and housed within the Body, and released. The breath of God is His Word. His Word is inhaled within and circulated throughout the Body to generate life support and energy for productivity. In addition, it is and again, is released back into the elements to provoke change towards cause and effect into the atmosphere.

HA

Intestine: is the portion of the alimentary canal extending from the stomach to the anus. It consists of two segments, the small intestine and the large intestine.

SA

Intestine: This is the area where spiritual discipline is practiced, and where anything that is unfavorable is eliminated. The environment of the spiritual intestine is like a threshing floor, which sifts through the bad in order to retain the good for strength, zeal, and creative energy. This is an environment where the process of purging for purification is at work. Notable character traits for this position would include the spiritually discerning, the prophet/prophetess, the revivalist, and biblical instructor.

HA

Liver: is the largest gland of the body that lies beneath the diaphragm in the upper right portion of the abdominal cavity. The liver which secretes bile, and is active in the formation of certain blood proteins, along with the metabolism of carbohydrates, fats, and proteins.

SA

Spiritual Liver: This is another area in which purging, sanctification, and purification take place. It is also the place where testing takes place. All of God's word is absorbed here, both the bitter and the sweet. It is then redistributed throughout the Body for power to illuminate or expose anything that is not holy or unacceptable by God. Here, the Lord's purging takes place by way of His judgments, testimonies, righteousness, and truth. This begins the process to reject and eliminate any inerrancy, unholy, false, or destructive activity within the Body of believers. Some who possess this trait operate in the true prophetic callings. They are no-nonsense believers who take God seriously for whatever He calls for of them. No matter how hard the saying is, they believe and do as He asks them. They strive to be submissive and obedient to the Lord. Their positional stance effects change through the Body of believers.

HA

Pancreas: is a lobulated gland without a capsule, extending from the duodenum to the spleen, and consists of a flattened head within the duodenal concavity. It has an elongated. three-sided body extending across the abdomen, and a tail that touches the spleen. It also secretes insulin and glucagons internally, along with pancreatic juice externally in the intestine.

SA

Pancreas: This member is one who operates within the apostolic doctrinal truths and practices of faith. Usually, they serve in this position from a priestly standpoint.

HA

Spleen: is a large, highly vascular lymphoid organ, lying in the human body to the left of the stomach, just below the diaphragm. The Spleen serves to store blood, disintegrate old blood cells, filter foreign substances from the blood, and produce lymphocytes.

SA

Spleen: This is characteristic of those who act in a representative capacity as watchmen to uphold apostolic doctrinal belief system.

HA

Appendix: is a pouch like vestigial process that extends from the lower end of the large intestine cavity.

SA

Spiritual Appendix: These are those who train and learn in the areas of higher biblical studies. They are instructors, trainers, and teachers of the Ecclesia.

HA

Kidneys: are a pair of organs in the dorsal region of the vertebrate abdominal cavity. They function to maintain proper water and electrolyte balance, regulate acid-base concentration, and filter the blood of metabolic wastes, which are then excreted as urine.

SA

Spiritual Kidneys: This is characteristic of word-based ministries, and those who practice water baptisms as a part of their initiation process in evangelistic ministry to new converts.

HA

Bladder: is a membranous sac that holds urinary fluid.

SA

Spiritual Bladder: a trait of those who hold positions of authority in the area of deliverance ministry. They heal the sick and cast out foul spirits.

HA

Pelvis: The pelvic arch or girdle, (Anat.) consists of two or more bony or cartilaginous pieces of the vertebrate skeleton to which the hind limbs are articulated. When fully ossified, the arch usually consists of three principal bones on each side: the ilium, ischium, and pubis, which are often closely united in the adult; working the innominate bone.

SA

Pelvis: This member is relative to those who support or aid the mobility and positional stance of the Body's framework. This is a typical trait and characteristic element of strength, utility, and unity to do battle or progress in movement.

HA

Reproductive System: has organs and tissues involved in the production and maturation of gametes. This tissues and organs work in unison towards the subsequent development as offspring.

Why are there no any sexual organs in the Spiritual Anatomy?

We are born again by way of the Spiritual Gonads of God. This process began with the immaculate conception of the virgin birth that had taken place through the Holy Spirit, who germinated the Seed of God into the virgin womb of Mary. Through the miracle of this pregnancy, she gave birth to Emmanuel (Christ Jesus), the Son of God. The only begotten Son who later died for the sins of the world. He stands as the redeeming source for all those who would believe on Him. By faith, believers are connected to the Spiritual Root to Jesus (Yeshua) the Christ.

Matthew 1:23, Behold, a virgin shall be with child, and shall bring forth a son, and they shall call his name Emmanuel, which being interpreted is, God with us. The spiritual body has no need for sexual organs, because there is no more male or female in the Body of Christ. Believers are adopted into the Family of God. The Heavenly Father's Holy Seal of approval is placed upon every believer of God's choosing. The Lord chose saved ones in Christ even before the foundations of the world began. God predestinated believers into His Glory, so that we would be set apart for Him, His purpose, and His Glory! The Lord foreordained and destined every believer to follow His plan, because it pleases Him. The Lord loves us all, and has set aside an royal inheritance for each of us that abide within him. Having revealed unto us by the Holy Spirit, the council of our purpose that's in accordance with His Will. (Eph. 1:5)

SA

Spiritual Reproductive System: There are no sexual organs within the Body of Christ. Yet, we reproduce spiritually by bringing others into the discipleship fold of Christ. When the true Gospel is preached and heard with active faith, the born-again experience occurs. The offspring are then shaped and molded into the image of Christ by the Holy Spirit.

However, there is a *Spiritual Womb*, but let's note the physical uterus (womb) of the woman.

HA

Womb: The womb (**uterus**) is a hollow, pear-shaped organ located in a woman's lower abdomen between the bladder and the rectum. The narrow, lower portion of the uterus is the cervix, the broader, upper part is the corpus. The corpus is made up of two layers of tissue.

In women of childbearing age, the inner layer of the uterus (endometrium) goes through a series of monthly changes known as the menstrual cycle. Each month, endometrial tissue grows and thickens in preparation to receive a fertilized egg. Menstruation occurs when this tissue is not used, disintegrates, and passes out through the vagina. The outer layer of the corpus (myometrium) is a muscular tissue that expands during pregnancy to hold the growing fetus, and contracts during labor to deliver the child.

The **Fallopian tubes**, one on each side, transport the egg from the ovary to the uterus (the womb). These tubes have small hair like projections called cilia on the cells of the lining. These tubal cilia are essential to the movement of the egg through the tube into the uterus. These tubes bear the name of the sixteenth-century Italian physician and anatomist Gabriele Falloppio.

The **Cervix** is the lower, narrow part of the uterus. It forms a canal that opens into the vagina.

The **Umbilical Cord:** is a flexible structure that gives passage to the umbilical arteries and veins. This connects the embryo or fetus to the placenta.

Spiritual Umbilical Cord: provides a link to the Holy Ghost through the preaching and teaching of the gospel. Where spiritual infants can be fed the Word of God in order to live and grow by. It is relative to providing a continuous eternal link to that which is essential for Spiritual Maturity and Development. This would also refer to maintaining a connection to the True Vine of God as the Source of all our needs and provisions within the spirit and body. I believe that the Spiritual Body never loses its divine connection to God as the source for all of its needs. I would also venture to say that there may be a Spiritual Maternal Twin to the Spiritual Body of Christ. That's linked to another Spiritual but not completely identical Twin counterpart, the Jewish Family of God . . . Both Maternal Families may possibly share from the same nurturing support of the True Vine. This Spiritual Umbilical Cord, being the True Vine, unifies all believers as one in the sight of the True God, the Heavenly Father.

From the Spiritual Umbilical Cord, flows the Blood of God's Justification, His Holy Water of Purification and His Spiritual DNA Identification. Also, God's Characteristics for godly behavior are imparted to us, through HIM being the Divine Source of Life.

The **Vagina** is the muscular canal extending from the cervix to the outside of the body. The word "vagina" is a Latin word meaning "a sheath or scabbard," into which one might slide and sheath a sword. The "sword" in the case of the anatomic vagina is presumably the penis.

SA

Spiritual Womb: is that special place of empowerment for giving birth to visions, dreams, hopes, and ministerial works. It is a place where passion and forbearance meet. Characteristic traits of those Christian believers operating within the spiritual matrix of the ***spiritual womb*** are: long-suffering, patience, love, meekness, resilience, and temperance. They are protective ministering agents that work to cover the growth and development of a newly formed ministry. Operational gifting placed here may care for the development of new covert believers, and support God inspired vision that is not yet born through manifestation into the physical realm.

What takes place within the Spiritual Womb?

The presence of our Father God is drawing nigh to the Church Body, while establishing truth and order within the *Spiritual Womb* to conceive the Word. When conditions are right, the word of God is spoken to breathe life for the germination of the Lord's promises. This works to brings light to the Vision that's cultivated within the spiritual womb. Thereby, perceiving and conceiving God's pre-constructed Destiny for manifestation of dreams and visions, according to His Promises over our lives. At the appointed time for giving birth to a godly inspired dream or vision: travail occurs, often prompting spiritual contractions to bring forth Destiny at its full term of development. That's when Manifestation of each spiritual reality is birthed into existence. The Spiritual Womb is also a place where the process of being born again occurs within each spiritual babe in Christ. Furthermore, it's a place where abandoned dreams, hopes and visions are revisited, reaffirmed and cultivated to full term. This will enable embarkment upon God's predetermined Destiny over wounded souls towards
reconciliation of His Purpose.

HA

The *Shoulder:* is the joint that connects the arm with the torso, and is also that part of the human body between the neck and upper arm.

SA

Spiritual Shoulders: are relative to those who carry the burdens of others, or uphold governmental obligations and standards. Also, these two members, both left and right, undergo and share tasks through their supportive effort to produce a collaborative work.

HA

Arms: are the upper limb portion of the human body, connecting the hand and wrist to the shoulder.

SA

Spiritual Arms: These are a combination of spiritual forearms, elbows, biceps, and triceps that work harmoniously to perform a task. These carry typical traits of those who represent Help ministries such as, people who support, nurture, bear up, transport, aid, and assist in mission work or kingdom building.

HA

Biceps: a muscle having two heads, as **a:** the large flexor muscle of the front of the upper arm. **b:** the large flexor muscle of the back of the upper leg.

Triceps: These are large extensor muscles that are situated along the back of the upper arms.

SA

Spiritual Biceps: typical of those who assist to compel or propel movement by adding strength to its counterparts towards collaborative and cooperative ministerial efforts.

Spiritual Triceps: characteristics are the same as spiritual biceps, in which they strengthen by the Power of the Holy Ghost. The spiritual triceps provide the added support in a inspired God breath ministerial effort.

HA

Forearms: the part of the arm between the elbow and the wrist. Also, they serve as the corresponding part in other vertebrates.

SA

Spiritual Forearms: This member's traits are typical to those who possess power and authority. They work as an extension to upholding righteousness, facilitating deliverance, fighting battles, and bearing gifts.

HA

Elbow: is the joint or bend of the arm between the forearm and the upper arm.

SA

Spiritual Elbow: positions and adjusts to join with and support its counterpart member by upholding authority, in order to implement the corresponding actions of a rule or plan.

HA

Wrist: is the joint between the hand and the forearm.

SA

Spiritual Wrist: As any joint does, it combines, cooperates, unites, and mutually works in conjunction with whatever it is attached to. In this case, either the left or right spiritual wrist—both serve in dual purposes to connect the hand to the forearm to fulfill a given task. This is a typical trait of those who work in Help ministries to coordinate, administrate, or partner up with others for Kingdom work.

HA

Hand: is the terminal part of the human arm, which is located below the forearm, and is used for grasping and holding. It consists of the wrist, palm, four fingers, and an opposable thumb.

SA

Spiritual Hands: these are expressionistic and godly instruments that represent: ***Faith, Love, Worship, Praise, Power, Ministry, Authority, Leadership, Championship, Administration, Delegation, Creativity, Development, Productivity, Orchestration***, and any other ***Handy Work of God.*** They operate to do the Will of God for the benefit of the Body of Christ. They're instructed to be instruments of productivity when to comes to performing kingdom work towards building spiritual growth, and the establishment of spiritual development. Spiritual hands address the need to reach out for the purpose of: demonstrating love, kindness, giving, sharing, support, receiving, healing, repairing, progression, building, and working.

In addition, they interdependently and harmoniously work together as a **spiritual hand** to bind what needs to be repaired or held together, and assists in the restoration of whatever is broken or lacking for the purpose of restoring order. This ministerial work serves to loose and release blessings through spiritual battle in prayer and intercession. Also, a spiritual hand will work together with the other **spiritual hand** to do battle, or to do the work of a builder. As **spiritual hands**, their position is to receive from God, and in return offer back to God in sacrificial worship, thanksgiving, and praise. The counterparts that make up the **Spiritual Anatomy's hands** also receive from God in order to sow seeds to plant, and reap a harvest for distribution to those in need.

HA

Palm: is the inner surface of the hand that extends from the wrist to the base of the fingers.

SA

Spiritual Palms: They are used to cup the anointing oil of the Holy Spirit. In particular, the left palm which represents servitude and the right palm, which represents the authority to release blessings, work together under the Lord's Headship. The palms are to receive blessings, tools, weapons, and instructions from God. They are also used as expressions of worship, submission, and prayer.

HA

Finger: is one of the five digits of the hand. Each digit can work independently among each of its counterparts, while also working harmoniously with one another to complete a task.

SA

Spiritual Fingers: are members that skillfully make or produce a work for God's glory. This operation of ministry is particularly relative to the right thumbs of Aaron's sons, which were anointed and consecrated by Moses putting blood on the tips of their thumbs. The act of anointing the right thumb was to receive the blessings of God for directing the hand for priestly service. God being the same yesterday, today, and forever more, never does change. In the anatomy of the Spiritual Body, the right thumb member is consecrated for priestly duties. Along with the other counterpart members serving as **spiritual fingers,** they also work interdependent to touch hurting souls, and to mend the broken-hearted. **Spiritual fingers** also are used for measuring, constructing, or for sizing up a task.

And he brought Aaron's sons, and Moses put of the blood upon the tip of their right ear, and upon the thumbs of their right hands, and upon the great toes of their right feet: and Moses sprinkled the blood upon the altar round about. (Lev. 8:24)

HA

Waist: is the part of the human trunk between the bottom of the rib cage and the pelvis.

SA

Celestial Waist: is the part of the Body that unites together in order to bridge and connect areas that strengthen, protect, and give progressive stamina. This is relative to those who work and assist in various ministries, including help, administrative, and support.

HA

Hip: the laterally projecting region of each side of the lower or posterior part of the mammalian trunk, which is formed by the lateral parts of the pelvis and upper part of the femur; together with the fleshy parts covering them.

SA

Hip: It enables weight to be equally distributed to either side of the Body for movement and support. If the hip is out of socket, the Body will limp. This is a typical characteristic of those who provide sound advice and doctrinal instruction. The spiritual hip is also relative to those who provide balanced judgment through prudent spiritual wisdom, knowledge, and understanding by the Word of God.

HA

Hip Joint: the ball-and-socket joint comprising the articulation between the femur and the hip bone.

SA

Hip Joint: Spiritual characteristic traits of this member enhances the movement that allows freedom to implement a cause. This spiritually balanced member will also align accordingly in order to coordinate unilaterally towards a positional course of action, while directing adjoining members towards a call for action.

HA

Buttocks: are the back portions of the hip, which form the fleshy parts on which a person sits.

SA

Celestial Backside: allows the Body to experience: spiritual rest, to sit in heavenly places, to assemble, and to be seated in the presence of God. This member will humbly posture itself in a position of worship or in acknowledgement of the fact that the Lord rules. Character traits of those operating in this member are in positions of priestly orders for: worship, praise, thanksgiving, and meditation.

HA

Leg: a limb used especially for supporting the body, and for walking.
a) Either of the two lower human limbs that extend from the top of the thigh to the foot, and especially the part between the knee and the ankle.

b) Any of the rather generalized segmental appendages of an arthropod used in walking and crawling.

SA

Legs: Each of these spiritual members are joined to a spiritual thigh, which in turn, is connected to the spiritual foot. They work together for the purpose of supporting the body and progressive motion. Both left and right spiritual lower limbs must move interdependently to allow the articulate expression of movement needed to transport the Body from one location to another. These characteristic traits are relative to members that pastor or shepherd converts. This is also typical of those operating in positions of leadership and Helps ministries.

HA

Thigh: is the portion of the human leg between the hip and the knee.

SA

Spiritual Thigh: This is a member that holds much honor, particularly the thigh on the right. Under the thigh is a place to take an oath or honor a request according to Jewish customs. The *spiritual thigh* also represents or carries a position of warfare. This honorable member must continually reflect righteous representation. This is typically where the sword on the spirit is readily placed to defend the principles and ordinates of the Kingdom of God. Characteristic traits of the member are relative to those who operate in the true prophetic gifting. Whatever the Lord says, they strive to obey His Decree.

HA

Knee: is the joint between the thigh and the lower leg, which formed by the articulation of the femur and the tibia. The interior of the knee is covered by the patella.

SA

Spiritual Knee n. pl. knees: a member that holds strength, vitality, and mobility. As the spiritual knees bend, they are in a position of prayer, humility, and submission before God. The *spiritual knees* bend in worship to God, while the tongue confesses allegiance to the Lord. Both spiritual knees enable the body to support the legs in walking, running, jumping, or standing up erect. Spiritual traits are exemplified through prayer, and humble expressions in grace, joy, love, faith, and godly actions. These characteristics are found in prayer warriors, intercessors, godly spiritual leaders, and priests. Their focus is to remain in fellowship with the Lord, through prayer, supplication and intercession.

HA

Calf: is the fleshy muscular back part of the human leg between the knee and ankle.

SA

Spiritual Calf Muscle: This member is typical of those who support ministerial efforts, while also assisting and encouraging others in a corresponding action of leadership. They also participate in priestly functions such as corporate prayer and worship. This ministerial component, along with prayer, works to provide assistance for the Body to stand. Typically, those with this characteristic act as a conduit towards prayer, praise, and worship orchestration.

HA

Ankle: is the joint formed by the articulation of the lower leg bones with the talus. The ankle connects the foot with the leg.

SA

***Spiritual Ankle** n. pl. ankles:* the members of both spiritual ankles allow fluid mobility of the feet and legs of the Spiritual Anatomy. They carry a spiritual trait of supporting a cause, missionary work, or ministry. These two delicate members enable the feet to take a positional stance in either direction to pivot, turn aside, leap, run forward, climb, step back, or to circumspectly walk straight.

HA

Foot: is the lower extremity of the vertebrate leg that is in direct contact with the ground in standing or walking.

SA

***Spiritual Foot** n. pl. feet:* This is where the Body represents spiritual authority, and establishes our righteous standpoint as believers over all principalities and spiritual wickedness in high places. Where God has placed the enemy underneath the Body of Christ's feet (as a footstool). The spiritual foot holds the enemy captive and under subjection. Under the spiritual foot is a place of conquest. Spiritual feet can oppose or reject what is not true or acceptable according to the standards of God's commands. The Body does not draw back unto perdition, but progressively moves forward to advance the kingdom of God. If either foot steps back, it's because there's a spiritual warning signal alerted. However, members of the Body can reevaluate a situation, and then repent. The message that's being transmitted to the feet would determine the mode of action governed. Humiliation and conviction are met at the feet.

SA

Spiritual Feet: This is the place where we proclaim the defeat of the enemy! Spiritual feet allow the senses of God's presence to be acknowledged and expressed. The Holy Spirit will also release the persuasive element of conviction to move sinners towards repentance. Both spiritual feet move circumspectly to take the Spiritual Anatomy of Christ to higher ground for battle. The spiritual feet are also capable of taking the Body to a place of peace, healing, and comfort. The Holy Spirit leads the Spiritual Anatomy with our spiritual feet. This sanctified Body should walk in God's ordered path of righteousness. These feet ultimately bring the Body before the Lord. The spiritual feet usher the Body to enter into His presence by way of the Blood of Jesus Christ, through thanksgiving and praise unto God. God establishes the feet, therefore allowing the Body to stand justified in Christ. Spiritual feet carry the Gospel that's prepared for those who will hear it. These members are spiritually sensitive to the holiness of God. The Body walks in reverence in the presence of God in the spirit submission and worship. There is victory in Jesus Christ through the power of His redemption and reconciliation. Overcoming power is obtained through Jesus Christ. Each member must learn to walk in the victory in order to experience its rewards. What does it mean to walk in victory? This would mean to move or stride, while endeavoring to progress forward in motion on a mission.

Character traits for this particular member are relative to believers that: preach, evangelize, heal the sick, minister deliverance, perform missionary work, and serve as peacemakers.

HA

Instep: This is an arched middle part of the human foot between the toes and the ankle.

SA

Spiritual Instep: This is a bridge support between the heel and toes that evenly distributes the weight of the Body on the foot. If the instep collapses, it will impede the ability of the foot to take proper steps. This part of the member supports all activity and involvement that promotes progress and productivity. The *spiritual instep* supports any advancement and territorial gain for the Kingdom of God. There is a certain level of love, joy, and enthusiasm that flows through this member, invoking the strength to encounter what waits ahead. The primary characteristic traits of these believers are as follows: burden carrying and operating within the call of evangelism.

HA

Heel: is the rounded posterior portion of the human foot under and behind the ankle.

SA

Spiritual Heels: The Body of Christ has the authority to subject its enemy to remain where he belongs; underneath the believers feet, to be treaded upon. After a victorious battle against the enemy, the Body will rest its foot upon the enemy's neck. It is evident of that God-given authority by the bruise that remains upon the devil's head by the heel. *Christ Jesus is the utmost example of this triumphant act, through His redemptive power that was performed on the Cross. Those who possess the characteristics of a spiritual heel are usually prayer warriors, intercessors, no-nonsense teachers and preachers, as well as spiritual watchmen.*

HA

Toe: Is one of the digits of the foot. The human foot usually consists of a total of five digits on each foot, and each one of those digits is equally significant in use.

SA

Spiritual Toes: Just as significance was placed on anointing the big right toes of Aaron's sons for God's direction concerning priestly duties and service, so is added honor placed upon the Spiritual Big Right Toe. The

SA right big toe member also holds a special position for disciplinary practices conducive towards good conduct for priestly duties and service.

Each spiritual toe member and its counterpart represent Kingdom Order. Its particular traits are of those who work to establish order in the areas of administration and leadership. Other traits would include the conquering of new territory for the advancement of the Kingdom with unity and strength. These members also have no outside or hidden agendas other than to maintain apostolic order, and to follow God's divine inspired design. All members of any part of the Church Body are equally and significantly important as a whole. Those who appear to have less significance shall receive more honors. In other words, there is balance within the Body of Christ. Each member is interdependent of one another to perform a work that glorifies God, but yet also remains codependent of God and His graces.

And as the toes of the feet were part of iron, and part of clay, so the kingdom shall be partly strong, and partly broken. (Dan. 2:42)

HA

Finger and Toe Nails: These are thin, horny, transparent plate coverings upon the upper surface of the end of a finger or toe.

SA

Spiritual Finger and Toe Nails: These are relative to sanctified coverings that protect the ministerial work of the spiritual fingers and toes. Spiritual toe nails are characteristic of members who protect the integrity of Kingdom ruler ship and its authority that comes from God. Spiritual finger nails strive to protect the orchestration and constructive work of Kingdom Building. They're financially empowered with kingship authority, which supports Kingdom building efforts to advance in dominion of territorial authority.

Each Member Is Important

The Body of Christ has many members operating in diverse operations and gifts. And although there are many members within the Body of Christ, we yet remain and function as one. From the greatest to the least, all members of the Body are vital throughout the entire body. Each member is different, and therefore each member's contribution is significantly valuable and important to the overall functionality of the Body. Every member has a spiritually designated place of order to operate from. These spiritual gifts are not gender-specific or driven. These gifts, members, and callings vary in offices and functions. Yet, they come from the same Spirit. If all members were the same, our abilities and functions would be limited. Therefore, while every member may have different functions and abilities within the Body, each function is instrumental towards the continual operation of the Body. There's no need to covet each other's position, but instead know that God has methodically placed everyone as it pleases Him. If we were all one member, such as a foot, we would not be a body, but a foot. Trying to position yourself to function outside of what you were called and skilled to do, could be troublesome to you, and disruptive to those surrounding you. This would impair the Body by causing imbalance or limited mobility. All members should remain fitly joined together. Consider, for example, a foot that is detached from the rest of the body would cease to function. If a body were missing a foot, it would hop rather than walk. Amen! We need each other, no matter how big or small the member is to the Body. We are not to despise the small things. God has given more honor toward those members who appear to be weak or insignificant. By God's design the lesser members are then capable of doing even greater works to His Glory.

For the body is not one member, but many. (1 Cor. 12:14)

1 Corinthians 12:16

And if the ear shall say, Because I am not the **eye**, I am not of the body; is it therefore not of the body?

1 Corinthians 12:17

If the whole body were an **eye**, where were the hearing? If the whole were hearing, where were the smelling?

1 Corinthians 12:21

And the **eye** cannot say unto the hand, I have no need of thee: nor again the head to the feet, I have no need of you.

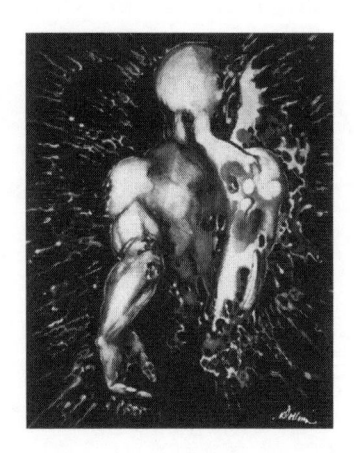

Foreign Elements that attack the body

HA

Toxicants: these are foreign elements that attack the body. Relating to a toxin or other poison. These toxins or poisons are capable of causing injury or death, especially by chemical means (poisonous).

SA

Spiritual Toxicants: These toxicants are intrusive agents that are comparable to a sinful or rebellious nature. Particularly, *spiritually demonic toxicants* that function as rebellious persuasions to spread their ungodly influence and disorderly conduct. Rebellion against God leads to destruction and death.

HA

Toxins: Are Collodial Proteinaceous poisonous substances that are a specific product of the metabolic activities of a living organism. Usually, they are very unstable and notably toxic, when introduced into the tissues. They are capable of inducing antibody formation.

SA

Spiritual Toxins: These are demonic toxins that try to invade the Spiritual Anatomy with crafty devices of deception. The source of these deadly and poisonous devices is the devil. These demonic toxins carry with them a measure of evil that petrifies and pollutes whatever it attaches to. The common device this enemy uses to attack the body is sin. Its purpose is to direct the individual towards sin, and to embrace everything that comes as a result of sin. This could result in rebellious acts of conduct, doubt, and unbelief in what God's word says. Doubt and Unbelief enter in through false doctrine and the evil influences of others. This can also happen as a result of believers taking their eyes off of Jesus and consequently, they lose sight of what He deems to be true.

HA

Germs

1. **Biology:** A small mass of protoplasm, cells, or living substances from which a new organism or one of its parts may develop.
2. The earliest form of an organism; a seed, bud, or spore.
3. A microorganism, especially a pathogen.
4. Something that may serve as the basis of further growth or development. The germ of a project.

SA

Germs: Spiritual Germs

1. **Biology of Spiritual Germs** can refer to a small mass of counterfeit cells that misrepresent themselves as something they are not. These cells can transform into a new organism, in which one or more of its parts may develop. These germs of deception are incapable of fulfilling a godly assignment or project. Instead, their purpose is to kill, steal, and destroy the work of God.
2. Similar to a diseased organism, the deceptive characteristics of these cells begins with a small seed, then takes root, and then buds as it continues on to multiply or spore. The Bible tells us that evil communication corrupts good manners. Sin is like leaven spreading to contaminate the entire measure of anything good for consumption. God cannot and will not accept anything that's tainted by sin! Any part that is sin-related must die first. If any redeemable portion remains, then it may be offered back to God as an acceptable sacrifice.
3. Just like a microorganism, sin may grow undetected by the natural eye, but if discerned from a godly perspective, sin can be detected from far off. The Spiritual Anatomy must remain on guard against

destructive and deadly diseases which try to shut the Body down from its life-giving Source. Demonic hindering devices strive to eradicate godly expressions, qualities, and characteristics.
4. Sin is a gradual process, although it starts as a small incident, it can grow if undetected. Eventually, it may evolve into a fully operating destructive force. It can start through a situation, person, or anything that acts as an element to grow or develop the hidden agenda of the enemies' methods of control.

HA

Pathogen: An agent that causes disease, specifically a living microorganism such as a bacterium or fungus. **Noun 1:** Pathogen—any disease-producing agent (especially a virus or bacterium or other microorganism).

SA

Spiritual Pathogen: A spiritual pathogenic agent could be a false prophet who propagates false doctrines through the telling of lies, in order to deceive many towards following their own hidden agendas. Their practices breed disease in any congregating unit. If continued manipulations occur and remain undetected, they will pave the way towards full-fledged disease infestation. This can be especially damaging for immature Christians, or individuals under the influence of occultist religious sects. These spiritual pathogens can start out as a small and insignificant spiritual microorganisms, such as spiritual bacterium or fungus. **Noun 1:** Spiritual Pathogen—any spiritual disease-producing agent (especially any unclean, uncircumcised hearted vassal of dishonorable mention) that serves to deplete the spiritual Body of its resources. These pathogens work to thwart the spiritual growth and life of the Body of believers. Moreover, these pathogens also strive to replicate other schisms of its kind.

HA

Microorganism—any organism of microscopic size

SA

Microorganism: A *Spiritual Microorganism* is any: spiritually tainted person, vassal, tool, element, device, or work of destructive properties. If members of the Body in Christ are not spiritually discerning, then spiritual microorganisms can remain undetected by the natural eye. Therefore, believers should be spiritually discerning when it comes to recognizing an oncoming attack to the Body. These elements must be dealt with immediately by prayer, fasting, and the full armor of God. This works to counteract the effects of these negative devices.

HA

Infectious Agent, Infective Agent: An infectious agent that is capable of producing infection.

SA

Infectious Agent, Infective Agent:

Spiritual Infectious Agent, Infective Agent: a sin agent capable of producing or spreading a hazardous infection within the Body of believers. Evil associations corrupt good protocol or behavior. This sin agent can come in the form of evil influences, rebellious conduct, individuals involved in occulted religious practices, false doctrines, or overall unbelief in the sovereignty of God.

HA

Bacteria: plural of bacterium (microbiology); single-celled or noncellular spherical or spiral or rod-shaped organisms that lack chlorophyll. These organisms reproduce through fission and fulfill the roles of either pathogens or biochemical properties.

SA

Spiritual Bacteria: plural of spiritual bacterium—the microbiological structure of a cell can resemble that of a Christian within the Anatomy of the Spiritual Body. However, Spiritual Bacteria are different in both outward and inward expressions. This works to produce characteristics and the overall nature of the anti-Christian. These anti-Christians reproduce their own kind of carnal church breeds that are unpleasing to God. Their influence and instruction come from the pit of hell. This is because their supply and connectivity come from the devil. Their support system is satanic, and is capable of splitting off and reproducing more of its kind. At times, they can creep in inconspicuously within a

Christian environment, and if not quickly terminated they can grow to reproduce many more of their kind, while remaining hidden in darkness. It's agenda evolves from the darkness that cuts off what is true and holy. And like natural bacteria, they can cause great harm as they sit in dark places and plant themselves, if they remain undiscovered. This is likened to a witch sitting in the congregation; someone who breeds strife, lies, division, prejudice, hatred, murder, and every other evil work.

HA

Resistant bacteria: bacteria that is unaffected by penicillin

SA

Spiritual Resistant bacteria: Their form may appear to be godly, but they deny the power thereof. This is typical of demonically oppressed or possessed individuals. They infiltrate and co-mingle with Christians. Spiritual resistant bacteria maneuvers to perpetrate its schemes to pose snares of attack against the Church. These sneaky antagonists can sit under godly-inspired teachings, and yet go on living unsaved lives. They're carrying out their own devilish agenda to penetrate the Spiritual Anatomy of believers in order to limit, debilitate, and corrupt the spiritually healthy environment of believers. They continue to spread their will to perform evil deeds and spread falsehoods that work to bring about destruction. These traits are typically expressed through: demon-filled entities such as witches, warlocks, false teachers, and anti-christs.

HA

Disease: an impairment of the normal state of the living animal, plant body, or any of the parts that interrupts or modifies the performance of the vital functions. It is typically manifested by distinguishing signs and symptoms. This is a response to various environmental factors (such as malnutrition, industrial hazards, or climate). These specific infective agents (worms, bacteria, and viruses), work to highlight inherent defects of the organism (genetic anomalies). Any combination of these diseases can work to bring harm to the Christian body. Sickness: A defective or unsound condition. ***Illness:*** Poor health resulting from disease of body or mind. Sickness.

Additional descriptions for disease are:

1. A pathological condition of a part, organ, or system of an organism, which can be brought about by various causes such as infection, genetic defect, or environmental stress. This is usually characterized by an identifiable group of signs or symptoms.

2. A condition or tendency of society that is regarded as abnormal and harmful.

SA

Spiritual Illness: This constitutes spiritual alienation or separation from God. This is caused by a lack of genuine fellowship with the Lord, and other believers due to one's life of sin, which results in the practicing of immoral conduct within the spirit of carnality and wickedness.

Affliction:
1. A state of great suffering and distress due to adversity
2. A condition of suffering or distress due to ill health
3. A cause of great suffering and distress
4. A state of pain, distress, grief, or misery
5. A state of mental or bodily pain, such as sickness, loss, calamity, or persecution

The suffering lies deep in the soul, and usually arises from some powerful cause, such as the loss of what is most dear—friends, health, etc. We do not speak of mere sickness or pain as "an affliction," though one who suffers from either is said to be afflicted. But deprivations of every kind, such as deafness, blindness, or the loss of limbs, etc., are called afflictions, which shows that the term applies particularly to prolonged sources of suffering. Sorrow and grief are very much alike in their meaning, but grief is the stronger term of the two, and usually denotes poignant mental suffering for some definite cause. Specifically, the grief one feels for the death of a dear friend. Sorrow is more reflective, and is tinged with regret, just like the misconduct of a child is looked upon with sorrow. Grief is often violent and demonstrative. Sorrow is usually deep and brooding. Distress implies extreme suffering, either bodily or mental.

Affliction in its higher stages denotes pain of a restless or agitating kind. Moreover, it is almost always related to some sort of struggle within the mind or body.

A. A condition or disease producing weakness.
   B. A failing or defect in a person's character.

Pain is a good indication that something is wrong and requires immediate attention for correction. Pain and affliction prompt us to change or react in a different manner.

   1. Ailment: Any sort of physical or mental disorder. Mild Illnesses are also applicable.
   2. Infirmity:

   A. Quality or state of being infirm. That which causes a lack of strength.
   B. a moral weakness or failing.

Synonyms for affliction: mishap, trouble, tribulation, calamity, catastrophe, and disaster. Affliction, adversity, misfortune, or trial all refer to an event or circumstance that is hard to bear.

HA

Cancer: any of various malignant neoplasms characterized by the proliferation of annalistic cells that tend to invade surrounding tissue and metastasize to new body sites. Any malignant growth or tumor caused by abnormal and uncontrolled cell division. This growth may spread to other parts of the body.

SA

Cancer: This is the most relative aspect of the sin's troubling effect on the individual. Sin will eventually begin to overwhelm and overtake the individual, leading to the depletion of spiritual resources, and ultimately resulting in the individual's downfall if not corrected. A member who is infected by spiritual cancer must become consecrated to the word and principles of God to live within the Body. Spiritual cancer must be isolated, and then finally removed from the Body of believers. Sin has neither the place or authority to remain within the Spiritual Body.

Transformation Factor

The Transformation Factor within the Spiritual Anatomy and Genetic Coding of the Human Anatomy:

HA

DNA: any of various nucleic acids that are usually the molecular basis of heredity. They are constructed by a double helix that is held together by hydrogen bonds between purine and pyrimidine bases. These project inward from two chains containing alternate links of deoxyribose and phosphate. In eukaryotes that are localized chiefly in cell nuclei they are called *deoxyribonucleic acid*.

SA

Spiritual DNA: God spoke unto His Seed of life, the divine infrastructure that determines our divine characteristic traits of who we are in the ***Body of Christ.*** In the ***Spiritual DNA***, we have a new nature, a ***godly nature*** that resembles our ***Heavenly Father's likeness.***

HA

RNA: any of various nucleic acids that contain ribose and uracil as structural components, and are associated with the control of cellular chemical activities.

SA

Spiritual RNA: from this comes the ***Plan of God***, which dictates who we are within the ***Body of Christ***. It's purpose is to carry this information throughout the Body of Christ, while also determining the length of our days upon the earth, and how those days will be lived out. The Spiritual Anatomy receives direction and divine orchestration from the ***spiritual brain***, which transfers impulses of information throughout the various ***spiritual transmitters*** and ***receptors*** throughout the ***Body***. ***Spiritual transmitters and receptors*** must remain resilient, yet sensitive to carrying or relaying God's information entirely throughout the Body with integrity, while maintaining pure motives.

HA

Double Helix: a helix or spiral consisting of two strands in the surface of a cylinder, that coil around its axis, *especially* the structural arrangement of DNA in space that consists of paired polynucleotide strands stabilized by cross-links between purine and pyrimidine bases.

SA

Spiritual Double Helix: This comes from the Mind and spoken word of God. It connects or pairs our ***spiritual genetic coding*** with the ***Spiritual DNA*** *of our Heavenly Father and Creator*, which causes us to inherently articulate and express the individual characteristic traits of our ***"Yeshua Ha Mashiach" Jesus Christ.***

HA

Alpha-helix: the coiled structural arrangement of many proteins consisting of a single chain of amino acids stabilized by hydrogen bonds.

SA

Spiritual Alpha-helix: This is relative to our genetic coding. Every member should innately strive to live their lives to the glory of God. Although, the Body of Christ is made up of many members; all with various functions or purposes, there remains a close connection to each genetically assigned purpose for executing and enabling it's uniquely driven impulses. This ultimately works to reflect godly characteristic traits unto the world.

HA

Immune reaction, immune response, immunologic response: a bodily defense reaction that recognizes an invading substance (an antigen such as a virus or fungus or bacteria or transplanted organ), and then produces antibodies specifically designed to fight against all that antigen.

Bioremediation: the act of treating waste or pollutants through the use of microorganisms (such as bacteria), that can break down the undesirable substances.

Acidophil, acidophilus—an organism that thrives in a relatively acidic environment.

SA

Spiritual immune reaction, Spiritual immune response, Spiritual immunologic response: These reactionary responses are likened to a spiritual lining or protection against sin, destructive impulses, or evil desires.

SA

Spiritual Bioremediation: this is war against the undesirable invasions of the enemy. ***Spiritual Bioremediation*** acts as an alarm system that deploys strategic anointed warriors that fight off the harmful effects of sin from the entire **Body of Christ**. It discerns and recognizes what belongs, while denying any further access to what does not belong within the body, therefore maintaining further access to ensure a healthy environment within the **Church Body**.

SA

Spiritual acidophil, acidophilus: This enables true **believers** not to be overtaken by the unbelieving, counterfeit believers, and false prophets who position themselves as **tares**. **The Anointing of God** maintains a proper balance for a healthy environment within the **Body of Christ**.

Be Good to the Body

Elements that promote health and healing to the human body:

Proper nutrition
Exercise
Reduced stress level

Elements that promote health and healing in the Spiritual Body:

Proper intake of the word of God
Faith, carried out by works, and adherence to the Will of God
Trust in God

Believers are responsible for taking care of their bodies. Also, believers are responsible for respecting and taking care of one another. Believers are expected to remain disciplined in the areas of promoting better health within their own physical body. This helps ensure that the Lord gets the best from each vassal. This is the believers reasonable service. Born again Believers are a part of the same Spiritual Anatomy. Love should be exhibited one towards the other in the Spirit. Spiritual members should exhibit a godly life style here on earth. Be a living testimony, and maintain a certain level of discipline over the deeds, inclinations, and desires of the flesh. For good spiritual health, we should maintain a proper intake of God's word. A spiritually balanced diet is properly digested. Spiritual food is absorbed for its nutritional value, and then circulated as power throughout the entire Spiritual Body.

Believers are a part of the same Spiritual Anatomy. Love should be exhibited towards each other in the Spirit of Christ. When other members are found to be hurting or weak, the strong believer(s) should show empathy one towards the other, since we are all part of the same body. From there, we will quickly respond to the needful repair or aide that comes to the healing of that ailing part of ourselves.

Becoming Spiritually Healthy

Believers must be God-centered in order to maintain proper balance in their lives. To maintain balance, you must continue to seek God through prayer, fasting, and submission to His word. Faith enables the Body to properly align itself with whatever God is calling for within the lives of each member. In order to submit to God, you must hear His voice, and recognize Him calling. You get to know God by faith; through hearing His preached word. Growth comes by having knowledge of Him by reading what the Bible says about Him. Therefore, you become accustomed to hearing His voice, communicating with Him through prayer, and listening earnestly for His responses. Through continued communication and fellowship with God, a relationship is developed. Furthermore, by applying what was sowed through the Spirit by the council of God will manifest results through an acquired experience with Him.

The Great Physician

If my people, which are called by my name, shall humble themselves, and pray, and seek my face, and turn from their wicked ways; then will I hear from heaven, and will forgive their sin, and will heal their land. (2 Chron. 7:14)

Jesus Christ is the Great Physician over the entire Church Body. Believers can be made spiritually and physically healthy, according to God's provisions for restoring wholeness to our lives. God has made provisions for spiritual and practical applications for health and wellness maintenance within both the physical and spiritual bodies. God's people must take unto consideration the physical health of their minds and bodies to enable a life of productivity and godly fulfillment. Also, with even a greater consideration, believers must come to respect God's Order, and then learn to operate within Kingdom principles to maintain spiritual health within the Body of Christ. Obedience to the Lord's directives will allow for the attainment of our spiritual goals, in addition to greater spiritual and natural fulfillment, as we experience the godly purpose of our lives. God's physical laws of nature and spiritual laws of Divine Order are not to be ignored. The Power of God's healing reaches every aspect of the physical, psychological, and spiritual needs, of all people. Furthemore, God will see to our overall development as well. Jesus Christ came to seek all of those who are in need of a physician. Of course, He was referring to Himself, who is able to heal both the outer and the inner man of a person. That would also include each and every one that remains outside of His healing virtue and restoration. If we are saved, then we are no longer under the curse of the Law of sickness and death. Born-again believers are privileged to the full measure of benefits to salvation, along with reconciliation, healing, and total restoration. Through God's reconciliation, we are redeemed and reconciled back to Him as our Father, God, and Lord. Through God's grace, saved believers are recovered and restored back to a life of completeness in Him. The redeemed are reunited to each other in the unity of Christ, because Jesus bore the wages of sin and death, by taking upon Himself the curse of the Law as He hung upon the Cross for us. Jesus, the Sacrificial Lamb, remained innocent and guiltless of any sin or wrong. Yet, He stood in our place, and paid the full penalty for the sins of the world. Christ became the propitiation for those who would trust and believe in Him. In Jesus' earthly ministry, He went about serving people and visiting them at the very point of their need. He offered Himself as: hope to the hopeless, food for the hungry, drink for the thirsty, covering for the naked, riches for the poor, sight for the blind, ear opener for deaf ears, healer to the sick and afflicted, uplifter to the weak, savior for the unsaved, restorer of the broken, filler for the empty, and guidance counselor to the lost. Christ stands as the compass for all those who are lost.

And Jesus went about all the cities and villages, teaching in their synagogues, and preaching the gospel of the kingdom, and healing every sickness and every disease among the people. (Matt. 9:35)

Conventional medicine or Alternative medicine?

Just a little forethought concerning natural remedies . . .

God heals sickness and brokenness, while also setting out to free captives from bondage. He has equipped and designed our body to repair and heal itself. There are also natural remedies that God has made available for us through healthy diets, natural herbs, and medicine. Both the physical and Spiritual anatomies are meant to live, and not die. These bodies don't know how to die on their own, in fact, they fight with all that is within them to live to the fullest. Much of the natural body's recovery occurs during our periods of rest or sleep. Rest is also a significant aspect of recovery in the Spiritual Anatomy through Christ. Jesus Christ is that Sabbath Rest which enables the Body to cease from doing any fleshly work.

Should we prefer conventional medicine to alternative medicine? The word *holism* was first coined in South Africa in the mid-1920s. "Natural healing" has its roots in the remedies used by tribes, settlers, and rural dwellers. It was the prevailing healing method utilized during the "heyday" of our ancestors. This is where I draw the line of precaution. We must be sure not to get caught up in some form of mystical or satanic-influenced chicanery by those associated with this kind of treatment. The devil is aware that these natural herbs do indeed carry some benefit. However, his primary job is to corrupt that which is true, and instead contort it to serve his own evil purposes. That's why there needs to be more Christians at the head of companies that perform business in the various fields of: science, medicine, agriculture, finance, communications, land development, and consumer resources. By doing so, we will better serve the Body of believers. So let's deal with all modalities with a level of discernment, all right? Now, as I have stated before, natural healing recognizes that the body is splendidly designed to resist disease and heal injuries. However, when disease or injury occurs, the first course of natural healing is to see what can be done to strengthen the body's natural resistance and healing powers. This is so the body can act against the disease or injury process. Since natural healing takes a slower, more organic approach, results are not to be expected with the frequency that some people expect with conventional medicine. This makes it less desirable than conventional and faster symptom suppressors. The flip side of this approach, is that the cure comes without the potentially damaging side effects that are so common within modern medicine. A natural healing orientation means that when you have a symptom, instead of reaching for the first available analgesic or rushing to the doctor to ask for treatment, you try the use of naturally occurring remedies and approaches first. Then, if all else fails, you resort to conventional medical attention.

Allopathic care refers to the treatment of the symptoms and not the disease. Most body functions act in a way to help rid the body of disease. Fever is caused when the body releases pyrogens to help fight infections (*Encyclopedia Index*, 2009). Over-the-counter medicines only help the comfort level of the patient. Diarrhea helps to rid the body of infections or bacteria causing the diarrhea. In some cases, antibiotics (the allopathic way) have been known to kill good bacteria (*The Body*, 2004)! Coughing helps the body to loosen mucus so the individual can remove it. Coughing also clears the passage way to remove dust particles or other foreign objects. Over-the-counter cough medicines are for the comfort of the users to help them sleep.

Naturopathy refers to a lifestyle that embraces healing the natural way. Naturopathy derives from a variety of sources; actually culminating into a coherent system that believes "only nature heals." When a patient visits a naturopathic doctor with symptoms previously not discovered, the naturopathic believes something inherent in the body is impeding a return to normal potential health (Graham, 2005).

The naturopath looks at the existence of the body, while using common sense to suggest healing practices and lifestyle changes that can lead to a better life. This could include more laughter, sunlight, rest (sleep), healthy dieting, nonsedentary life, finding a life purpose, while also learning to love and be loved. The biggest challenge for the naturopath is to help the patient discover the proper balance between all of these. (Graham, 2005).

Did you know that homeopathic medicines are manufactured using exact pharmaceutical methods in agreement with the Homeopathic Pharmacopoeia of the United States (HPUS)? Well, unlike herbals and dietary supplements, homeopathic medicine manufacturing must meet strict FDA guidelines for strength, quality, purity, and labeling. Each homeopathic medicine is obtained from a precise and controlled procedure of successive homeopathic "dilutions." This process dilutes the original organic solution, and then transforms it into a therapeutically active medicine.

Some may ask, how safe are homeopathic medicines?

Several scientific and clinical studies have been conducted to determine the safety of homeopathic and other medicines. These studies are conducted in accordance with all FDA guidelines. For two hundred years, homeopathy has proven to be safe and effective in treating a wide array of illnesses and injuries—physical, mental, and emotional.

Homeopathic medicines are extremely safe, and carry no risk of side effects. Moreover, none come with the risk of addiction. Even if one takes the wrong remedy for the wrong condition, it will do no harm. This makes homeopathic remedies especially well-suited for babies and the elderly. Because of their inherent safety, they are frequently used for treatment of self-diagnosable conditions. They provide a natural and affordable alternative to over-the-counter conventional drugs.

http://www.coliccalm.com/baby_infant_newborn_articles/homeopathy.html
www.banyanbotanicals.com

Encyclopedia Index. (2009). Fever. Retrieved July 7, 2009, from
http://www.drhull.com/EncyMaster/F/fever.html
Graham, P. (Dec. 2005). What is Naturopathy? *Positive Health,* Issue 118, pp. 16-17, 2 p.
Retrieved July 7, 2009, from EBSCOhost database.
The Body. (2004). Diarrhea. Retrieved July 7, 2009, from
http://www.thebody.com/content/art6030.html

Scientific Theories

Here are some scientific theories that inadvertently prove that there is a God!

Scientists and cosmologists theorize at the ideas of Einstein's M theory. The String theory states that the existence of all things are orchestrated by a force that seems to operate like musical notes being played. My response to that analytical documentation is: Yes, there is a Chief Orchestrate, who when He speaks it's like a melodic rippling effect of a God-spoken word. This scientific theory also says that all particles of matter are not really particles, but strings of matter. And my response to that hypothesis is: All things are connected to their true Source, which is the God of everything that is, was, and ever will. I'm careful not to say ever will be, because everything already is in Him, our Creator. Even things we've haven't seen yet!

There is research surrounding the pseudoscientific theory of Biorhythms, which is the study about life rhythms or biological cycles of physiological, emotional, and intellectual well-being or prowess. The studies have not been able to prove them to have any predictive power or scientific fact. This is the world's attempt to try and tap into the spiritual realm to manipulate and influence immaterial information to deceive and control lives. There is spiritual wickedness in high places. These negative evil forces move about in the air, and even seek to possess the unsaved, in order to control their minds and bodies as evil instruments. They are demonic spirits that move to and fro, seeking who they may destroy, by working to pervert the truth into a lie. These demonic forces are able to see some things that are in the future, and their job is to try and distort, control, and manipulate the minds and lives of others. There is space between distance and time. When God says that it's finished, He really means it. Just because our eyes cannot physically see what's already done, doesn't mean that it hasn't. We that still reside in our physical bodies are caught in time. God works outside of time, yet orders things to happen inside of time according to synchronization of universal events, people, and works. Astronomers say that we are looking back in time, when it comes to looking beyond earth's atmosphere. The nearest two stars—Alpha and Proxima Centauri—are a little over four light-years away, meaning that the light from those stars takes over four years to reach us. So we are actually seeing a distant image that is no longer present. We are viewing distant images from a past tense perspective. However, the Lord God has synchronized time and space to come together in moments of time. He has calibrated events to occur in our present. The Lord's work is done; we are just catching up to the reality of His already finished work. The devil knows that his time is short, and he wants to destroy and deceive all of mankind.

Researchers may detect biorhythms in space that relate to sound. But, I would not trust their subjective response to that information which was obtained without spiritually discerning the nature of the information derived. Because they cannot pinpoint the nature and origin of this immaterial information. They are not trustworthy unless they are able to point to God as being the Originator of Life for this particular theory. So, the research cannot be trusted as a creditable source for that opinion. Therefore, I would have to draw from the current findings, and form my own opinion concerning the information. The research may factually point to God as being the Originator of Life, but don't expect the devil to be righteous and tell the truth. The devil knows God is real, and trembles at the very hearing of the name of Jesus. Don't count on this evil force to change, instead only expect the enemy to continually lie, steal, kill, and destroy. Scientific research is very helpful at times, and some findings are dependable and factually sound. This information just proves that scientists fall short when it comes to Spiritual matters. Unless, he or she happens to be a scientist that's a born-again believer. God provides individuals with the intellect to concoct ingenious ideas within both the saved and unsaved. Only a person who believes in God, and will give Him the credit for the ability to think, and reason. And to joyfully honor Him with praise for having the proper functioning of our limbs, along with the very precious gift of life together. Christians need to learn how to use a combination of our God-given abilities of intellect, wisdom, and spiritual understanding. Believers must believers must learn to balance spirituality alongside their natural intellect. If they don't, they will become either spiritually stupid, too ethereal-minded, and will remain constantly out of touch with God's natural order of things. This sort of disorder causes a person to violate God's natural laws and boundaries around them. We are in the world, but as the redeemed, we are not of the world. However, believers still have to live here until God calls us, or comes to get us! The devil does not care if we fall to either extreme, just as long as we don't get it right! This primarily occurs, because we are out of harmonization with God's Truth.

Another theory is the 11th dimension idea. This is the idea that there is a universe parallel to ours. Theorizing that one universe is enclosed by a membrane, and the outer universe is covered by another brain. My response to that big idea is: There may be another universe, but God's heavenly throne and Kingdom order is on the other side of it. God is beyond even what is known to be the "11th Dimension." Also, the big bang theory is another big idea, where they stumble at the facts. My response to this well-known lie is: Yes! there was a great explosion in the universe, but God set it off! That which is factual proves that there is a Master Creator from which all things exist, and have their being. There may have been a big bang, but the bang came from the climatic effects when God said Let there be, and that word did produce matter and life. Furthermore, I will surmise that God the Father is the Source of where all scientific formulation emanates. God is the Author of all divine principles, and the Orchestrator of all universal laws. He's the Beginning and Ending of every process.

We should show forth the reflection of God and in the oneness of a triune earthly manifestation of His offspring in the third dimension.

Kingdom Come Mentality

Let the Church be ready for the coming of our Lord and Savior Jesus Christ! For, it is in our undying anticipation that we continue to find the pulse of the Heart of God. He is to remain the Source of our life-giving Force. The Church must remain aware that the Kingdom is about Him and His coming. When the King is expected to arrive at any given moment; the citizens are to prepare for Him with joy, reverence, and godly order. Our Kingdom comprises the unification of many colonies and tribes that are designed to bring about corporate worship and shared fellowship among believers. This is what God desires for his believers. His people are to form a support system of communal partnership and sharing. Being synergized together toward kingdom-building efforts. We are to know exactly where we stand in Christ, specifically, in terms of our gifts and callings. This also includes our ability to use god-given tools to build within the Kingdom. This includes utilizing the armor by which we have been equipped to fight. Our joint efforts are to be concentrated towards winning over the lost by representing Christ-Likeness unto the earth, in addition to worshipping and honoring the Lord in holiness. We are to recognize His Authority as being the Supreme Headship. He is above all things important. No other agendas can enter into the scope outside of pleasing Him. So, let us run with patience to our high calling. Let's leave no work undone!

    This book is to present a systematic schematic of how we as the Body can possibly relate, assist, and work together as a whole Spiritual Nation. Our citizenship is in a heavenly country under the governmental overseer of God. We are a people having the same spiritual origin and sharing the same language. There ought not to be any separation among us; instead we are to be perfectly joined together by having the same mind, and utilizing the same spiritual judgment. Despite our diversity, there is still solidarity among each godly member. Our thoughts are made perfect as we connect to the Mind of Christ, and align ourselves to the Word of God. Combining our cognitive thoughts and intellect to collectively serve God with unified minds and hearts. When we function as One Mind, and in one accord, there is nothing that would be impossible for us.

If believers function as one people speaking the same language, then nothing they have begun to do or plan to do will be impossible for them. Genesis 11: 6-7.

For we are labourers together with God: ye are God's husbandry, ye are God's building. According to the grace of God which is given unto me, as a wise masterbuilder, I have laid the foundation, and another buildeth thereon. But let every man take heed how he buildeth thereupon. For other foundation can no man lay than that is laid, which is Jesus Christ. (1Cor. 3:9-11)

Our Armor of God

How do we as believers remain in the faith of God? Let us be strong in the power of God's might, so that we be fully equipped in God's provisions that enable us to fight. This doesn't pertain to natural fighting, but the fight that rages against spiritual wickedness on the terrestrial and celestial levels. Therefore, we must put on and keep on the armor of salvation. This goes along with the salvation and power of God's rightness, faith, truth, and peace. Let us cast down all vain imaginations and evil suggestions; resist the devil, and submit to the Lord, by allowing God's word to speak to our hearts towards facilitating change. We must remain connected and directed by the word of God; knowing that our life is hidden with Christ in God. Believers are protected in God through His Plan of Redemption. The old life, and the recompense of it is covered by the Blood of Christ. Now, we have new life through His sacrifice. Once saved, Born-again believers are then baptized into the Body of Christ. Through this process of salvation, the saved are then fitly joined together with other members as one whole Body. It is important to watch and pray continually in the spirit of intercession for the Body of Christ.

Put on the whole armor of God, that ye may be able to stand against the wiles of the devil. For our wrestling is not against flesh and blood, but against the principalities, against the powers, against the world-rulers of this darkness, against the spiritual hosts of wickedness in the heavenly places
(Eph. 6:11-12)

Coming Together of the Body

The Body must come together and reconfigure itself to receive each other as an intricate part of its whole self. Having a renewed mind that's conducive to operating in the mind set of Kingdom Dynamics, will enable the blessings of God to flow within and outside the Body. Respecting Kingdom Order under the direction of the Holy Spirit of God, will allow each member to become spiritually balanced in God. At that point of transition, the Body will experience a transfiguration that's so amazing, it will cause a total identity change resembling the likeness of Christ.

So when Aaron and all the children of Israel saw Moses, behold, the skin of his face shone, and they were afraid to come near him. (Exod. 34:30)

Assuredly, I say to you, there are some standing here who shall not taste death till they see the son of Man coming in His kingdom. Now after six days Jesus took Peter, James, and John his brother, brought them up on a high mountain by themselves, and was transfigured before them: and his face did shine like the sun, and his raiment was white as the light. (Matt. 16:28, 17:1-2)

Who am I? Why don't you remember? Our spirits are eternal and existed even before the world was formed, while we were yet one in our Heavenly Father's bosom. He had predestinated us to be reconnected and reunited to Him, so that we would glorify Him on earth. The Master Architect of the entire universe programmed our spiritual DNA to fit together in a helix of life giving expressions . . . Our spiritual genetics are so uniquely designed and coded, that no other person or thing can fit into that personally hollowed place reserved for each other. A place that's perfectly aligned to accept the reunion of our spiritual traits to form a specific component that produces God-related characteristics and attributes within the spiritual anatomy of the Church. Together, we are rejoined within the Body to exhibit the qualities and virtues of the Living Christ. We are bone of His bone and flesh of His flesh, yet without sin, through the Blood of Jesus Christ. The Holy Spirit is the Living Force that allows the Body to be governed, to move, and have its being. Our spirits are reunited with His Spirit, and our mind is transformed and conformed to the Mind of Christ. Together, we all become One! Jesus Christ is the Head of the Church Body; His Face is the ensample of who God is, and the word that He speaks to the Body will instruct and direct. This allows His Body to retain mobility and expressions of God-inspired actions.

*For as the body is one, and hath many members, and all the members of that one body, being many, are one body: so also is **Christ**.* (1 Cor. 12:12)

*Now ye are the body of **Christ**, and members in particular.* (1 Cor. 12:27)

Unity within the Body

Let there be no division within the Body of Christ. Although many members, yet we are one Body.

Believers are to be longsuffering and forgiving of each other's faults by the Love of God. We are to make every effort to stay united together in the Spirit, while striving to remain connected in peace. Godly people are as one within the Body of Christ. Unity in Christ is the culmination of one Body and one Spirit, even as believers share one hope for the future. There is one Lord, one faith, one baptism. There is one God and Father, who is over all and in all, while living through all. Christian believers are all baptized into the Body of Christ through the water baptism in the Name of Jesus, the "Yahsua Ha Mashiach" in Hebrew. Believers are brought into oneness by the Power of the Holy Ghost. Believers come together through the unity of the same doctrinal faith and the belief in Jesus Christ as Lord, God, and Savior!

Now these are the gifts Christ gave to the church: the apostles, the prophets, the evangelists, and the pastors and teachers. Their responsibility is to equip God's people to do his work and build up the church, the body of Christ. This will continue until we all come to such unity in our faith and knowledge of God's Son that we will be mature in the Lord, measuring up to the full and complete standard of Christ. (Eph. 4:11-13)

Instead, we will speak the truth in love, growing in every way more and more like Christ, who is the head of his body, the church. He makes the whole body fit together perfectly. As each part does its own special work, it helps the other parts grow, so that the whole body is healthy and growing and full of love. (Eph. 4:15, 16)

We must unite together in God's love for the common good of the entire Body of Christ. We are to serve within the mind of Christ while remaining within the Holy Spirit that helps the Body to operate in unity. This works to produce a God-effect throughout the earthly realm. This is until the fulfillment of his Kingdom fully comes.

Unbelievers cannot fellowship within the Body of Christ. Therefore, the unsaved cannot participate as one within the Body. They can only remain outside the Body, except they be drawn by God's Spirit through the repentance of sin and faith in Jesus Christ.

Let all the house of Israel therefore know assuredly, that God hath made him both Lord and Christ, this Jesus whom ye crucified. Now when they heard this, they were pricked in their heart, and said unto Peter and the rest of the apostles, Brethren, what shall we do? And Peter said unto them, Repent ye, and be baptized every one of you in the name of Jesus Christ unto the remission of your sins; and ye shall receive the gift of the Holy Spirit. For to you is the promise, and to your children, and to all that are afar off, even as many as the Lord our God shall call unto him. (Acts 2:36-39)

The Magnificent Work of the Body of Christ

Believers should embrace a gift as God reveals it to you in the Spirit. A gift that the Lord has already prepared for you to receive. As He reveals the gift to you, prompting you to ask and give thanks for it by faith, even though it's already yours. There are both spiritual and natural gifts given for us to behold and be a blessing with. People should acknowledge their various gifts, and give God the glory for them. Ultimately God has given the supreme Gift in Christ and by His Spirit.

Spiritual gifts are not gender-specific. They do not come with certain social, educational, or economic stipulations. The only requirement is that you believe in the Gospel of Jesus Christ. You must be born again to receive spiritual gifts from the Holy Spirit.

There is neither Jew nor Greek, there is neither bond nor free, there is neither male nor female: for ye are all one in Christ Jesus. (Gal. 3:28)

It is important to show the world that Christians truly care. This is done by sharing surplus resources in outreach ministries and missions, while offering wraparound services and provisions to all who are in need. Furthermore, we extend to recover lost souls, heal the sick and afflicted, feed the hungered, clothe the naked, and to set the captives free. These are opportunities for us to reflect the love of God to the world, and share the Gospel of Jesus Christ towards winning lost souls.

It's important to extend love among each member in the Body of Christ. Not forsaking the coming together with other believers in the Body of Christ. Be attentive to the needs and concerns that are within our global corporate Church fellowship.

There is a great work for the members of Christ to do. This work is to be performed with the cooperation of each member, as we unite together for the purpose of loving our God, and being an expressed image and manifestation of His Glory. This is for the purpose of proclaiming the Gospel of Jesus Christ, and compelling others to follow Him in discipleship. Our primary aim should be to live a life that glorifies Christ. We cannot identify with Him without knowing who He is. As we identify with Him through faith and obedience through His Word; we embrace His Name. We identify with him through the act of submission and baptism. This process begins by allowing the Holy Spirit to convert our lives into His. We throw out the old carnal life, and begin to walk in the newness of Life with God's purpose. As we take on His nature and character, we can evoke godly influence that reflects Kingdom Order in the name of Jesus. God has given His people the power to subdue the earth; and He has given His angels the charge of watching over us, and preparing our way in peace. It's in the name of Jesus that we obtain the right, authority, and power to take dominion, while also serving as ambassadors of God. Then we can fulfill our destiny . . .

Fit within the Pattern after the Order of Melchizedeck

Christ Jesus, who is also Yeshua Ha Mashiach; the Messiah whom the Jews look for. He is the first Order and High Priest by which all born again believers must pattern their lives after. God is calling for His people of all nations to be as One within His Pattern. Where is my positioning within the spiritual matrix of the Church Body? I am called first to be a child of God, and then to worship and praise Him. My gifts enable me to teach others, write, administer, and help as needed. I believe that my position as a member of the Body would be in the Spiritual Stomach. My callings are determined, qualified, and endorsed by God. Your calling will coincide with your gifts. The value of your character will constitute the level by which your gift will operate. Testing and trials help to develop character, while putting you in touch with your passion. God will not call you to do something without anointing and equipping you with the gift(s) to handle the calling. So, wait on your gift(s) and be pliable to the Lord, while He molds your character. One of my specific gifts would be my calling to share through teaching. My work is to help spread the word of God. This position enables me to have godly influence that can be carried throughout the Body. Some of my other gifts would include, but are not limited to: discernment, word of wisdom, counseling, creativity, networking, hospitality, singing... I cannot truly put an end to the level of my gifts and callings, because there is always room for continued growth and discovery, as God broadens my territory to expand my spiritual and personal development. I cannot completely define who I am; only God can. And in Him, there are no limits!

By reading the following Bible verses, you will find that Christ is forever our high priest, and was also called of God to be the high priest after the order of Melchizedec. Also written as Melchizedeck. The name Melchizedeck means king of righteousness. *Bible verses:* Gen 14:18-20; Ps. 110:4; Heb. 5:6, 7:17, 21

I have begun to discover what my calling is in God, by first: submitting unto the Lord, enquiring of Him through prayer and supplication, reading the word of God, and committing to develop a strong and intimate personal relationship with God. Furthermore, my utmost desire is to please Him in every area of my life. I also seek to acknowledge him in everything that I do, while also taking advantage of every opportunity to serve others. It is a wonderful thing for the most high God to allow me the opportunity to discover gifts that combine my abilities with his heavenly guidance. This also includes spending personal time with Him. He has touched and anointed me to be a blessing. My soul, mind, and me are synonymous. He has awakened my soul to be sensitive of opportunities that avail themselves concerning the sharing of my gifts with others. It is in the sharing my gifts with others, that I exercise my calling. Every believer of God is called to be a blessing. My purpose and course of fulfillment is to: love the Lord my God with all that I am and do, to glorify God, and to lovingly serve others in honor of Him.

In all things showing thyself a pattern of good works: in doctrine showing incorruptness, gravity, sincerity. (Titus 2:7)

It's Better to Please God

Genesis, chapter 4, verses 2-8 tell us that God preferred Abel's offering over Cain's. Abel, being a keeper of sheep, presented a blood offering or meat offering. And Cain, being a tiller of the ground, presented a grain offering. Now, what would make God prefer one over the other? Well let's see . . . God is righteous, so His judgment is unwavering. This means there's no room for error. He simply does all things right. Ask yourself . . . what do you think would displease Him? Would performing all that He requires or asks of you do it? I seriously doubt that would offend Him. You're doing all you can to follow His commands. I think that He would be pleased for the effort. By God knowing all things, He already knows the limitations of your capabilities. Would He ask you to do something that you could not do? Well that depends . . . Since He already knows the end results of all things, He is either preparing us through instruction, to reprove or teach us a lesson, or to help us complete the task for His glory. Would disobedience or rebellion make Him displeased? Absolutely yes! Would doing things our own way, without considering or preferring His will over ours, do it? Oh well! Somebody's asking for a chastening there! Being deceitful, arrogant, or prideful toward God—these are definite ways to offend Him. When God tells you to do something His way; instead of choosing your own way, take His instructions to lead you. His way will always bring you better results. However, if you unknowingly do the wrong thing, out of innocence; God would lovingly correct us, if we adhere and submit to His Loving Hand of correction.

All right . . . now back to the initial question: Why was Abel's offering preferred over Cain's? I believe that both Abel and Cain knew what God's requirements were. I also know that Adam taught his family how to prepare sacrificial instruments of worship to a Holy God. I also believe that God told each of them their responsibilities, His required preferences, and the order of presentations to please Him. Blood sacrifice or grain sacrifice, which is better? The answer is whatever God is asking for at the specific time, is what He should get. Now, if we refer back to the Levitic law concerning the different types of sacrificial offerings that were required, you would find that certain types of offerings were required for certain types of sacrifices—whether there be bullocks, lambs, or doves to fulfill a certain blood or burnt sacrificial requirement. This could be extended to include grain, wheat, sliver, gold, or any other type of monetary offering. Each sacrificial offering was received according to what the nature of the sin was, and according to whatever the priest was requesting for atonement or worship. In other words, the people had to follow instructions to know what was acceptable at the appropriate time. You couldn't bring a spotted, blemished, or inferior offering either. It would be rejected.

Let's present the right offering according to God's request.

Did you know that a wheat offering can also be considered as a meat offering? I was surprised to read it in the following passage from the Bible:

> *And Ornan said unto David, Take it to thee, and let my lord the king do that which is good in his eyes: lo, I give thee the oxen also for burnt **offerings**, and the threshing instruments for wood, and the **wheat** for the meat **offering**; I give it all.* (1 Chron. 21:23)

I believe that God was requiring a blood offering for that season of Cain and Abel's offering. But instead, Cain just presented what he had, rather than joining together with his brother in fellowship, and sharing in giving a more favorable offering. That being a specially required blood offering. God was looking for the Blood. Amen, the Bible tells us that without the shedding of blood, there is no remission of sins. The Lord also said the following to the Israelites: "If I see the Blood, then I will pass over." This was a type and shadow of things to come. Amen, Jesus is the true Passover, and the true Sacrificial Lamb.

Now, because of the perpetual nature of sin that proceeded Adam's disobedience, rebellion was then able to enter into the heart of Cain. Therefore, Cain developed a disposition of pride, when presenting his offering by not fully considering God's Will over his own. Furthermore, his offering was not well received by God. Rather than repenting, his nature of pride continued until he slew his brother Abel. Consequently,

Cain became the first murderer when he killed his brother Abel, out of jealousy. Note that what was done in secret came to light. God sees and knows all things. Nothing is done without Him knowing about it. So, He asked for Cain's confession concerning the evil deed that was done. Instead, Cain replied, "Am I my brother's keeper?" Wrong answer! We are our brother's keeper. This was an example of how a bad deed, followed by a bad response, ultimately produced a bad result. God saw the error of Cain's ways, and made available the opportunity for Cain to acknowledge and confess the wrongdoing. This would help to assuage his guilt and engage his will to repent. Instead, Cain's response was solely selfish in considering no one but himself.

Here begins Cain's judgment:

> *When thou tillest the ground, it shall not henceforth yield unto thee her strength; a fugitive and a vagabond shalt thou be in the earth.*
> (Gen. 4:12)

Cain repents and cries out in the next verse:
And Cain said unto the Lord, My punishment is greater than I can bear. (Gen. 4:13)

> *Behold, thou hast driven me out this day from the face of the earth; and from thy face shall I be hid; and I shall be a fugitive and a vagabond in the earth; and it shall come to pass, that every one that findeth me shall slay me.* (Gen. 4:14)

Amen, Cain was fearing for his life. Now God's mercy is shown to Cain right here in Genesis, chapter 4, verse 15:

> *And the Lord said unto him, Therefore whosoever slayeth Cain, vengeance shall be taken on him sevenfold. And the Lord set a mark upon Cain, lest any finding him should kill him.*

God's mercy is evident! He still desires to cover and protect His children. Even today, our Lord has provided a seal of His approval for us by way of the covering of the Holy Spirit, which guides, protects, and leads us unto all truth.

God Is Love

God is the source from which all love emanates. For God is Love.

Amen . . . God gives His gift of love to us without the need of repayment. However, if we are truly thankful for all He has done, we will humbly submit to Him, and show our appreciation by serving him with a grateful attitude through our words, actions, and deeds. We can never repay the Lord for all of the blessings that He has so freely enriched and enhanced our lives with. But we ought to say thank you. We ought to show gratitude. And we ought to simply love God back, because He first loved us. So often, we take God's goodness for granted, as though we deserve His grace and blessings. When on the other hand, we deserve to be put to death because of our rebellious and sinful nature. But in spite of ourselves, God saw fit to offer us the gift of His only begotten Son, to take our place in receiving the rightful punishment for sin.

*And this hope will not lead to disappointment. For we know how dearly God **loves** us, because he has given us the Holy Spirit to fill our hearts with his **love**.* (Rom. 5:5)

Presenting ourselves as an instrument of God's Love is not an option, but is in fact, a prerequisite to exhibiting godly character. When we show love in the spirit of Christ out of an emotional and spiritual mind set; not looking for any for any sort of repayment, we show that this is a free gift that we are willing to give. To demonstrate Christ-like love, we must begin by receiving God's love, in order to return it. Because through his giving, we are able to receive the benefits of this great love. We then grow to love Him in return because of Who He is. God proved His love toward us by sending Christ to die for us, while we were yet sinners (Rom. 5:8).

> 1 Corinthians 13:13 tells us that "three things will last forever—faith, hope, and **love**—and the greatest of these is **love**."

Although faith, hope, and love will always remain with us, the one that stands out among the others is love. All things are possible, because of God's love concerning all of His creation. He's the source from which all things do exist. Without Him, there is nothing, and there would be no need for faith or hope. We have hope because of what God has promised. His promises are sure to be brought to fruition. He can do anything but fail.

There's tangible evidence that God loves us, by way of the Holy Spirit that He gives believers. The evidence of speaking in other tongues as the Spirit gives utterance to those who will receive of His free gift of Salvation. Amen . . . It is through this event that we are given the assurance that we are truly born again, because of the Love of God that is shed abroad in our hearts. His Love in us also allows us to love each other in brotherly love. Our love for God and each other, will testify that we are disciples of Christ. Through the initial gift of Christ comes many gifts and blessings. Jesus is the gift that keeps on giving. Praise God! The Lord's blessings are abundant!

> *Prophecy and speaking in unknown languages [Or in tongues.] and special knowledge will become useless. But **love** will last forever!* (1 Cor. 13:8)

There are times when speaking in tongues will not be perceived as one of the acceptable or beneficial gifts. But, I would like to further clarify that there will come a time when the unsaved will not be able to receive the gift because of doubt, and the inability to comprehend the origin and nature of the gift. This gift, along with other spiritual gifts will appear as bizarre to the unlearned. However, the unsaved may possibly be more receptive to our efforts of exhibiting God's love toward them. Every believer is drawn to repentance by the cords of God's Love. Therefore, showing Christian love will always be useful in reaching others, and drawing them in to receiving the loving presence of God through faith. Furthermore, we should also demonstrate love toward other Christians, and not despise or forbid their practice of exercising the use of prophecy and speaking in unknown tongues. Because these gifts are always useful in ministry toward building ourselves and others in faith. The blessing of praying and speaking in other tongues, comes in the knowledge of knowing that we are speaking mysteries, while we are being enlightened through communication with God.

*[Tongues and Prophecy] Let **love** be your highest goal! But you should also desire the special abilities the Spirit gives—especially the ability to prophesy.* (1 Cor. 14:1)

*And do everything with **love**.* (1 Cor. 16:14)

Now, let's examine another version of love, which is called Eros. Eros is a Greek word derived from a mythological god of love. This is the same word from which we get the word *erotica;* meaning to have sexual attraction and physical desire toward others.

Loosely practicing any form of eroticism can open you to a world of addictive behaviors and lustful cravings, which will lead individuals to resort to great lengths to gratify these carnal desires. Erotic desires gravitate towards what it craves, through the sensory perceptions associated with the sinful nature of carnal thinking. Carnal desires are never totally satisfied; these being human mental interpretations of receptive values of physical senses, that are charged through the eyes, ears, touch, and taste. There is a difference between physical and spiritual senses. Spiritual senses are in tune to what is intangible, or that which is eternal or heavenly. And of course, erotic sensations are compelled by that which is earthly and fleshly. God has given us a natural desire to have companionship. The problem comes when you search outside of His Will to fulfill these desires. This is when you stop considering God's directives and provisions for a mate. Some physical attractions can be divinely appointed—for example, the love that Isaac and Rebecca had for each other. Abraham, Isaac's father, had faith to pray and believe that the Lord would send an angel to prepare the way to find a wife for the son of his kindred. His servant who was sent, also swore to Abraham concerning the matter. And Rebecca was busy being a good servant, and ended up with her blessing. Both Isaac and Rebecca liked each other at first sight. Genesis 24:67 tells us that she became his wife, he loved her, and Isaac was comforted after his mother's death.

Ruth 2:4-15 tell us how Ruth submitted unto the Lord, and in doing as Naomi had suggested, had obtained grace in the sight of Boaz. Upon their meeting, Boaz was enamored by Ruth. You should learn that when you are open and submit to the things of God, then the Lord will lead you to your blessings . . .

Alright, single people out there, I know it's difficult to remain under the reproving hand of the Lord, as He is shaping us into persons that continually exhibit godly character. But, you must remain under His correction and refinement to get through the process of growth. This process can be further delayed by ignoring what is true and Holy, and instead going after wants in spite of the warnings, much like Samson did when he fell for Delilah. Later, Samson repented and cried out to God to give him another chance to do His will. God did bless Samson once again to destroy the enemy. Amen, it's still not too late; the Lord can put anyone back onto the proper course towards being blessed.

You must enquire of the Lord before making choices. It's a dangerous thing to be conditioned and stimulated by outside appearances, ungodly influences, ungodly influences, inward tendencies, and proclivities. But, we should aspire to understand the spiritual intent of a thing through godly spiritual discernment. Seek God first! Avoid any potential snares that could bring us down into a state of oppression, depression, or worst-case scenario: total destruction. Furthermore, do not be unequally yoked with anyone. This can apply to those considering a partner in marriage or business. Mainly I'm speaking of believers being unequally yoked with unsaved people. However, sometimes you can be unequally yoked with another believer who has not yet reached the same level of spiritual maturity as you have. This also can prove to be a problem for the believer, because the word of God asks how can two walk together, except they agree. You can't successfully work together with another; unless you both are able to submit to the Lord, commit to the purpose, or come to a common agreement to perform a given task . . .

So let's avoid the pitfall from the start!

Can two walk together, except they be agreed? (Amos 3:3)

God Made Sex to Be Good

Now, I must explain that physical attraction and its expression through physical stimulation is good when it remains under the guideline in which God had originally intended for us to both benefit and enjoy. This pertains to the genders of both male and female; in holy matrimony only. In plain language, you must be married first. That's a marriage union between one man and one woman with each other in a holy covenant with God. Again, this covenant is upon one married partnership at a time, otherwise, it's called adultery. The reason why I emphasize holy matrimony is because the world is trying to devise another order of marriage. Even if the world passes laws that legalize same-sex marriage, or people marrying their animals, God, the ultimate judge, says NOT SO! God will not accept the solicitation of those types of marriages, because it is ungodly, and out of His order. Nor is it good to be drawn in a sexual attraction that involves male to male relations, female to female relations, or human to animal relations. These warnings also extend to other sexual relations that are deemed to be ungodly combinations. God is against sexual perversion: Note the following:

Adultery—according to Deuteronomy 22:22-29, the sin of adultery carried a great recompense of reward, by putting both guilty parties to death; unless one of the parties involved was deemed to be the victim of rape or sexual assault.

Prostitution—as noted in Deuteronomy 23:17, according to the Levitic laws prostitution was not permitted, and was considered an abomination unto the Lord. This restriction also extended to include the act of homosexuality. Abomination means an evil thing. In that time period, they referred to homosexuals as being sodomites, hence derives the word *sodomize*.

Incest—as mentioned in Lev. 18:6-18, and throughout the entire chapter, stands as another unholy sexual act the brings great displeasure to the Lord. Throughout the chapter, readers will see God's extreme displeasure with this act. The Lord speaks against all manners of incest, and all the different variable combinations of sexual contact between close kinfolks or blood-related persons within any given family. Other areas of sexual misconduct and unseemliness are covered in the same chapter.

Some of you might say, "Well, that was the Old Testament." As we search further, this message concerning God's dislike of abominable acts is carried over to the New Testament, which also deals with the following:

Homosexuality—it's mentioned in Romans 1:26 and 27.

Mankind with beast—this problem is being dealt with in Deuteronomy 27:21.

Not to say that I am judging this sort of conduct, but judgment came upon those who practiced sexual misconduct or other abominable acts. The word of God plainly states how the Lord Himself regards this—when He deals with defilement in Lev. 18:22-28—and how those practicing such acts would be cut off or destroyed.

Again, this message of destruction carries over to the New Testament for those who refuse to believe. Destruction of the flesh will come even upon those in the congregation of believers who practice fornication. They will fall into judgment according to 1 Corinthians 5:1-5. The bible warns us that some will be delivered into the hands of Satan for the destruction of the flesh, so that the spirit might be saved. Amen . . . This is a sobering thought saints! Finally, judgment comes in the form of death, as mentioned in Leviticus 20:13-16.

There's more scripture that deals with the destruction of those who fall into homosexuality, incest, and bestiality. God is not at the forefront of these illicit desires, nor does He endorse such conduct. I am not placing judgment; only God can do that. His judgment will be based upon His word. Sin is sin, and we must call it what it is. For we have all come short of the glory of God. But we should not attempt to cover our sins. Only the Lord Jesus can cleanse us from sin and iniquity through the atonement of His Blood, saving grace, and mercy. We should acknowledge our wrong, repent for it, and desire change. I hope I was clear on that and made it plain. Amen.

I must remind you that the remedy for turning aside God's wrath and judgment, is to repent. To repent means to have godly sorrow for any wrongdoing or omission. Then we are to possess an active will to turn away or to cease from doing whatever wrong was done. Acts 26:20 tells us that there are works befitting repentance. So, we are not to just say that we've repented, because repentance requires an active engagement of one's will in order to have a changed mind, change of heart, and a complete turn from being offensive to God. By having a change of heart or mind, we are then deciding to do things God's way. The act of repentance requires you to engage the mind and emotions, while denying self-will. However, I realize that these are hard sayings. I realize that some will not understand, believe, or even submit to the truth that was told. But it still had to be said after all. I also know that there are many who will understand, and echo the word of God, while receiving the truth of His word. Then you will eat both the bitter and the sweet of it. God will richly bless you for your submission and belief according to His word.

Why should believers continue to marry, stay married, or remain celibate until the rapture of the Body of Christ?

It is a matter of remaining an undefiled member. The marital bed of a husband and wife is sanctified, undefiled, and blessed. Apostle Paul said that if you have the gift of celibacy, then it's better to stay unmarried in order to be more focused on doing things that please God. Not everyone will marry, or produce children. So, whether you're married, single, with or without children, you should learn to be content in the Lord. However, marriage in the Church, has its place of glorifying God as well. It is in marriage that we present a type of illustration that resembles the marital bond between Jesus Christ and the Church Body (Rev. 21:9). Therefore, there are blessings and benefits to receiving a godly orchestrated love for a mate.

In this section, I will serve you some sweet morsels of hope and encouragement. Here are some guidelines in which sexual and physical desire remains a blessing, as long as it is within the confines of God's precepts and ordinances.

The Bible tells us that sexual love was meant to be good and holy (Gen. 1:27-28; Gen.2:24-25). So, I will reiterate that it's to be in marriage only. I know that this is a difficult saying because it's a very pleasurable practice among both married and single people. But single people out there: You have to get married to properly receive the benefits. And the Bible tells us that it's better to marry than to burn. Now according to Proverbs 5:15-20, sexual pleasure should be within the framework of marriage only.

Sexual love within the confines of marriage is an expression of love, and it is so beautifully described in Songs of Solomon 1:12-15 and chapter 3, along with 1-5 of the same book.

Within the martial relationship, there should be shared mutual responsibility upon the husband and wife to submit their bodies to each other sexually and affectionately in love toward each other. Married couples are not to withhold their own bodies from each other; but for the sake of a limited time to devote to fasting through the permission of their spouse . . . according to 1 Corinthians 7:3-5.

God first created man; thereafter, God created woman, bringing her forth out of the man. Both male and female were created in His own image . . . creating them both in His own image . . . (Gen. 1:27).

Also, sharing of the marital covenant vows, is to be followed by the coming together of the husband and wife sexually to consummate the union of marriage, before entering into lifelong companionship. (Gen. 2:23-25).

For helpfulness . . . *God said that it is not good for man to be alone* (Gen. 2:18); so, He created woman to help meet the need. I believe that the need extends beyond companionship alone.

One aspect of physical attraction, is the need to procreate or to produce offspring for the Lord . . . (Gen. 4:1).

A husband and wife are to desire to be physically connected sexually to each other alone, and are to gratify each other's sexual needs within the marriage . . . (Prov. 5:17-19).

Physical and marital love began between Adam and Eve, and was ordained by the divine preparation and arrangement of God.

And out of the ground the Lord God formed every beast of the field, and every fowl of the air; and brought them unto Adam to see what he would call them: and whatsoever Adam called every living creature, that was the name thereof. (Gen. 2:19)

And Adam gave names to all cattle, and to the fowl of the air, and to every beast of the field; but for Adam there was not found a help meet for him. (Gen. 2:20)

And the Lord God caused a deep sleep to fall upon Adam, and he slept: and he took one of his ribs, and closed up the flesh instead thereof; (Gen. 2:21)

And Adam said, This is now bone of my bones, and flesh of my flesh: she shall be called Woman, because she was taken out of Man. (Gen. 2:23)

Now, if we were to study the proper marital love relationship between a man and a woman, including those of you who are married or thinking about getting married, we should refer to *Ephesians 5:22-33*. This will give you insight on the proper order of marital love, and its relative expression of exhibiting the love of Christ toward the Church body.

Single people should be open to the leading of the Lord. Don't be closed minded, by only expecting God to send a mate that matches the snapshot that's in your head. Well, the chances are your God-sent mate might not resemble what you thought of at first. I suggest that you remain prayerful through the process of seeking a companion. But find contentment in whatever status that you are in. If you are unmarried, take the advantage of having the spare time to develop a close relationship with God the Father, and getting to know who you are in Christ. Be busy about your Heavenly Father's business, and don't waste a lot of time feeling sorry for yourself. Instead, your time will be well spent doing things that produce godly effectiveness in your life, as well as being a positive influence in the lives of others. Providence might have it that God may be preparing you and your mate to be better suited to receive each other, while you're growing in maturity and blessing others.

However, instead of being in the business to pursue a mate, we should instead be in the business of seeking God's Face through prayer and supplication, and then will the Hand of His blessings be upon us. We have to enquire of the Lord to assist us in our choices. We don't want to be just stimulated by outside appearances or inward inclinations; but desire to understand what the spiritual intent of a thing is. We can't just go after what looks good, or talks good, or what even feels good.

Learn to seek God first! There should be a proper perspective concerning sexual love. God has ordained that sexual love is only beneficial within the confines of a marital relationship. However, I could not ignore the fact that there are a lot of frustrated Christian singles out there who are trying to hold on for the appropriate conditions prior to entering marital bliss. I hear your pain, and understand your plight. More importantly, God knows and understands your true desire. But, He says seek first the Kingdom of God, and His righteousness, then all of these things shall be added onto you. Amen, the Lord wants us to prioritize properly, by putting things in there proper order. Understand that everything we have ever needed and rightfully desired, is within His provision and care. If there's a relationship with Him, adherence to His directive, and submission to His Headship: then He will guide us by his hand. He will protect and watch over us, just as any king would protect and watch over his people. Don't just seek after the blessings of the Lord without first acknowledging Him as the God who holds the plan of all destiny. Beloved, be aware that there is preparation before marriage, which starts way ahead of any ceremonial obligations. Respect and honor God's authority. If everything is all about self gratification, then there's no readiness for entering into marriage. Because marriage is not about just you, but it is also about revealing a representation of the Bride of Christ, by demonstrating love and submission to the spouse. The act of submission would also mean: to be in obedience, to be in compliance with, defer to, or give in to, or surrender to. So, if you are still selfish and self-centered, you're not yet ready, because marriage will require a daily sacrifice of yourself. This love has balance, otherwise, it's not godly love. Love does not seek to blame for an offense. It is forgiving toward others, as well as yourself. When love is appropriate, it assumes responsibility to be accountable in finding a resolve. Here are a few verses from the Bible, so that you will know that marriage is a life of blessings and sacrificial love. As you read Ephesians 5:21-33, I will interject on these verses. The Bible says the following, beginning at verse 21.

subjecting yourselves one to another in the fear of Christ. (verse 21)

This verse is referring to both husband and wife; they are to surrender to each other's love and show respect.

Wives, be in subjection unto your own husbands, as unto the Lord. (verse 22)

Amen. Wives, you are to submit to your own husband, and no other man should operate as headship over you. Only Christ should hold chief headship over you both as members of the Church body. Then the husband is head over his wife, and after that order comes spiritual leadership such as pastors, other godly leadership, and finally public authorities. Children must honor and obey their parents, and respect their elders to ensure a prosperous life. Obey those who have rule over you, and respect the law of the land that is enforced. But God is the final authority over all laws, and Husbandmen, to Church Bride of Christ.

For the husband is the head of the wife, and Christ also is the head of the church, being himself the saviour of the body. (verse 23)

All right, this mentions the divine order within a marriage, or the rank of authority. The husband must represent Christ as protector, and willingly sacrifice himself to care for her.

The position of authority in marriage begins with Christ, second the husband, and third comes the wife; their children are subject to each of them.

But as the church is subject to Christ, so let the wives also be to their husbands in everything. Christ holds authority over the body of Christian believers; likewise wives must also respect and consider their own husbands in all that she does. (verse 24)

Wives inquire of and acknowledge God, your heavenly Father. The husband is to be your earthly representative of the Lord. Acknowledge the Lord concerning any vital decision-making, and follow the lead of your husband as he follows after Christ.

Wives are to respect and surrender unto her husband in everything; as there is no one who holds greater earthly authority than his under God.

Husbands, love your wives, even as Christ also loved the church, and gave himself up for it; (verse 25)

Amen . . . Now husbands, don't expect to receive this kind of respect unless you are representing headship in the spirit of Christ. Otherwise, you are showing yourself to be operating out of order, and lacking proper balance. You will come across as a control freak. Both the husband and wife should submit to each other in the Spirit of Christ. How can you create balance without first proper consideration, direction, and application from the Lord? So husbands, as the scriptures show, you must love your wives as Christ loved the Church by loving it so much, that He sacrificed and laid His life down for it. That was the most selfless act that you could ever imagine. So brothers, you have a major work ahead of you.

The husband is also to be responsible for representing a priestly position in his household. He ought to speak the word of God, be committed to prayer on behalf of himself, and his family for divine direction. The husband should also act as a priest, and petition the Lord's blessings and covering to be upon himself, as well as his wife and children.

that he might present the church to himself a glorious church, not having spot or wrinkle or any such thing; but that it should be holy and without blemish. (verse 27)

The marital relationship that a husband exhibits toward his wife is to be a reflection of the love relationship and glorious work of God upon the church. Husband and wife are to submit to one another.

Even so ought husbands also to love their own wives as their own bodies. He that loveth his own wife loveth himself: (verse 28) *for no man ever hated his own flesh; but nourisheth and cherisheth it, even as Christ also the church;* (verse 29) *because we are members of his body.* (verse 30)

Amen, these verses enables us to know that the man is to truly see himself as being one with his wife, just as we are all members of the same body of Christ, and the Church. The husband should love, value, nurture, and protect his wife as Christ does the Church. This is an awesome responsibility for the man to show love, while being the godly example of a Christ-centered lifestyle. If a man is acting in this capacity, then the woman should have no problem submitting over to that type of authority. It does not mean that he is a pushover. Women, you can't be domineering over him; in doing so, you are out of order. Women, you can destroy your household by acting disorderly, unless your husband is acting like the devil, and bringing total chaos into the home. In such case, he throws the family unit out of order himself, therefore leaving the wife to fend for herself. This forces her to get into self-preservation or provisionary mode through prayer, until there is godly intervention for divine order. Women, you are equal, but you are the weaker vassal too. In other words, you are to be your husband's balanced counterpart. You are not the same, but you both complement each other, and work together to be a better self-portrait of God's image.

For this cause shall a man leave his father and mother, and shall cleave to his wife; and the two shall become one flesh. (verse 31)

Bless the Lord! For the sake of the love covenant that a man shares with his wife now, they must work to form their own household, with the husband standing as authority over the household. They are no longer one flesh, nor are they joined or subjected to their parents any further.

This mystery is great: but I speak in regard of Christ and of the church. (verse 32)

Praise God. Yes, this is truly a mystery because we see through a glass darkly. And eyes have not seen, nor have ears heard, the riches that God has in store for all who believe. So beloved, we are to just keep our hands to the plow and do the work of the Lord, even though we don't fully understand how the marriage between a man and a woman can be a reflection of how Christ is married to the Church. We will see and understand the fullness of this truth when the Lord comes back for His Bride the Church. On that glorious day when we are called up to meet Him in the air. Then we will be like Him, because we will see Him as He is.

Nevertheless do ye also severally love each one his own wife even as himself; and let the wife see that she fear her husband. (verse 33)

So the bottom line, is men must strongly love your wives, and wives should respect your husbands; as they are the head of the household.

Ephesians 5:22 reads as follows: *"Wives, be in subjection unto your own husbands, as unto the Lord."*

The union of husband and wife presents such an opportunity to represent a visual illustration of the marital Spirit-to-Spirit bond relationship between Christ and the Church.

Man and woman were created to be one within the Wholeness of God. They were identified as a part of each other; being formed by God in the completeness of the Whole. God is whole and complete in and of Himself. He does not need anyone else. We need to be one with God to be complete. Whether we are married or single, God is the One who completes us.

Male and female created he them; and blessed them, and called their name Adam, in the day when they were created. (Gen. 5:2)

The marital relationship that a husband exhibits toward his wife, is to be a reflection of the love relationship and glorious work of God upon the Church Body. Godly marriages are to be that kind of example, or that type of model. But the responsibility does not rest solely upon the woman to be make all of the necessary sacrifices. The man must also stand before God as a Christlike representative, who displays the sacrificial love that exemplifies selfless living. Husbands, you know when you are living and acting godly.

As someone mentioned, "What about wives who must submit to unsaved or abusive husbands? How does that work?" I know that various interpretations are floating around that seek to keep women in bondage. But you are to submit to a husband who loves you and desires to be with you in peace. Women, this is such a blessing if you already have a godly husband; don't blow it by being controlling and disrespectful! Instead, love him and cherish him! You can even submit to a husband who is unsaved, if he desires to be with you . . . In other words, women, you can always respect a man who loves and cares for you, unsaved or saved, as long as he is not abusive in his position as headship over your body. If either the husband or wife is abusive one towards the other, carrying on in adultery, or tearing the other down, then divorce stands as a viable option. Furthermore, if the husband or wife positions themselves as the devil incarnate, then that spouse is not showing a true desire to love, or remain in the marriage. God does not favor divorce. However, due to spiritually devastating reasons, divorce is permitted in this case. Just know that the idea of divorce should be considered a last resort, not an easy way to bail out. Divorce is permitted for lack of love, unfaithfulness, or hardness of heart (Mark 10:4). There must be repentance and reconciliation, if there's to be recovery within the relationship or marriage.

God's Judgment! How Should We Perceive It?

Deuteronomy 6:1

Now these are the commandments, the statutes, and the judgments, which the Lord your God commanded to teach you, that ye might do them in the land whither ye go to possess it.

Amen, the Lord wants His commandments, statutes, and judgments to be taught and handed down to others under our tutelage, so that all might follow thereafter. And, in acknowledgement of these ordinates, we're to take them as God-given directives to be adhered to wherever we go, and to appropriately implement them into whatever we do . . . His commandments are rules or orders that we must abide by. His statutes are directives to apply or live by. As we learn by way of God's judgments, we receive clarity or focus to what His word declares is right . . . It is a barometer for what is true. This judgment is balanced through His infinite Wisdom, and is justifiably and infallibly correct. Besides that, He's God, and it's His prerogative to do whatever He pleases. His judgments are designed to establish order, punish evil, chasten, manifest His righteousness, correct us, and warn us. Typical aspects of God's judgment can come is the form of: physical destruction, material loss, spiritual blindness, and eternal destruction. Ways to avoid punishment are as follows: Acknowledge and confess your faults to God, repent, and turn away from excising any sinful offense. Humble yourself by submitting to God's correction and guidance, while acknowledging and owning up to our wrongdoing. And of course, we can commit to prayer, and then submit to His will in our lives . . . For the Lord tells us the following in 2 Chronicles 7:14: *"If My people, who are called by My name, shall humble themselves, pray, seek, crave, and require of necessity My face and turn from **their** wicked ways, then will I hear from heaven, forgive **their** sin, and **heal their** land."*

Now, this verse sounds like this is a pretty good prerequisite for turning away God's wrath, don't you think? OK, moving forward . . . Now, what are His Statutes? Statutes are boundaries, limitations, or warnings established by God's Law. His limitations are not set in place to make our lives miserable, but instead work to protect us, and enhance our quality of life here on earth. They promote wellness and correctness. Now, the law of sin and death will do just that: present to us a life of misery and eternal damnation in hell. Oops! Did I just use the "H" word? It's a word that's very seldom used today. People are offended by it, and think that it's taboo to mention, but since it is a place of warning from eternal judgment, I'll just say it for the hell of it. It does exist! It's a real place! Evangelists such as me, who talk about hell, do not get invited to speak as often as those who speak only entertaining messages. They even get fewer offerings too! But I've got to stick with the truth. Think about this for a moment . . . If there was a fiery pitch somewhere that you were about to step into at any given moment, wouldn't you like to be warned of the apparent danger that lies ahead of you? Or would you like someone to say "step right up!" There's a great adventure that awaits you while you're on your way. Here are some goodies for you that are available for your fun and comfort as you take your trip"? Or would you prefer to be told the truth, even though it was a hard saying, in order to spare you from total destruction? Or better yet, would you prefer a loud warning or a soft whisper? Selah . . . But I'm not surprised by the lack in numbers in following the truth, because the Bible tells me this:

> *Enter through the narrow gate; for wide is the gate and spacious and broad is the way that leads away to destruction, and many are those who are entering through it.* (Matt. 7:13)

As I sought the Lord, and inquired of Him, and grew in spiritual maturity, I pondered upon this translation taken from the King James Bible:

> *Enter ye in at the strait gate: for wide is the gate, and broad is the way, that leadedth to destruction, and many there be which go in*

> *thereat: Because strait is the gate, and narrow is the way, which leadeth unto life, and few there be that find it.* (Matt. 7:13, 14)

Amen. I won't expound much on these verses right now because this is a whole other topical study. But, I will say that not many will follow after the narrow path of righteousness, and few will be able to find the path that leads unto eternal life. What a sobering verse. I realize that I've expressed some subject matters that present a hard saying, but I must present to you the word as God gives me to give to you . . .

If I was presenting to some another Gospel, then they would love me; but since I tell you the truth, not many will accept me as an instrument of God to serve . . . But I thank God that there is a remnant out there who will hear His voice and receive His blessings . . .

How Should We Express Love?

God is Love and the source of all that animates from His Love. What are the practical applications of love as we know it? You may wonder if the reactive response to love is an emotion, or a spiritual response that does not require reciprocity, or if Love is a free gift. I would say that it is a combination of all three: an emotion, a spiritual response without any need of repayment, and a free gift that costs the giver what cannot be repaid. However, love can prompt a relational reciprocal response of continuous reciprocity. Love is more than a notion. You can not love without commitment, and you can't commit without love . . . So where do you begin? By faith, we learn to love through God's love for us . . . As I love you, I'm loving Him back through you . . .

Different phases of love:
    The first would be Agape, which is a Greek word rendered as love and charity. It is descriptive of God. For the scripture tells us that God is Love.

*The one who does not **love** does not know **God**, for [1 John 4:7, 16] **God is love**.*
(1 John 4:8)

God's love is unconditional and sacrificial toward others. Because He first loved us, even though we are undeserving of His Love. We are shown the embodiment of what pure Love is expressed in Christ. And we are admonished to highly regard His love as sacred and true. The main objective should be to love God who has given His Gift. Upon acceptance of His Gift: out comes Love, and from the Love flows life that is everlasting. Out of this life-giving force flows continuous blessings upon our lives. These blessings continually and abundantly overflow towards others to bless.

And He said to him, [AC] You Shall Love The Lord Your God With All Your Heart, And With All Your Soul, And With All Your Mind. (Matt. 22:37)

The Lord deserves the very best from every child of God. Fortify your love towards Him, and maximize your every potential to flourish throughout every element of your being. The Lord is worthy of such admiration, reverence, and love. In fact, you can not love Him enough. Commit to love the Lord and submit to Him. Then communication with Him will become more meaningful, and your relationship will become more intimate with Him. The true self originates in Him, and is discovered through Christ. So when we love God, we learn to love our true self. We were created by God our Creator for Himself, and for His good pleasure.

Love for neighbor:

This is the great and foremost commandment. (Matt. 22:38)
The second is like it, [AD] You shall love your neighbor as yourself (Matt. 22:39)

The Bible specifically indicates the primary commandment is to first love the Lord God with all our might, and secondly, we are commanded to love each other, as ourselves. If we love ourselves, then we will also love our counterparts, which is an extension of who we are in Christ. Loving others is relevant to how we love ourselves, and this is the evidence of our love for God. Because, how can we say that we love God, who is a Spirit, and do not love others that are in the image of Him that we can see?

Also verse 40 says that *"[AE] On these two commandments depend the whole Law and the Prophets."*

Amen. Verse 40 tells us that by having godly love, we fulfill the Law. And our religion is made pure by possessing Christian love, which is sacrificial in giving, with no underhanded motives.

Here's a description of what having Agape love is all about: Let's read 1 Corinthians, chapter 13, verses 5, 6, 9-12 from the King James version:

Doth not behave itself unseemly, seeketh not her own, is not easily provoked, thinketh no evil; (verse 5)

Love operates in the spirit of meekness or self-control. Love does not compel you to think about doing harm toward others. Neither are you quick to react in a rebellious or condescending manner.

Rejoiceth not in iniquity, but rejoiceth in the truth; (verse 6)

Don't rejoice when you, or others fall into sin. But rejoice in God's word, which is true.

For we know in part, and we prophesy in part. (verse 9)

Because not everything is revealed at this point, and we only know in part what is to become; even as we perceive through faith.

But when that which is perfect is come, then that which is in part shall be done away. (verse 10)

When we appear before Him who is prefect, there will be no need to rely on faith; we shall be like Him, because we'll see Him as He is.

When I was a child, I spake as a child, I understood as a child, I thought as a child: but when I became a man, I put away childish things. (verse 11)

There is an appropriate time to be a babe in Christ, and there's a predetermined time to grow in the knowledge of the truth, and put away acting immaturely . . . .

For now we see through a glass, darkly; but then face to face: now I know in part; but then shall I know even as also I am known. (verse 12)

Even though we see the things of God vaguely, there will be a time when our blinders shall be totally removed, as we are before our Living Savior face to face.

Love for family:

Here is an example of even a slave's love for his family; he chooses to remain with his family instead of being released as a free man.

*And if the servant shall plainly say, I **love** my master, my wife, and my children; I will not go out free: Then his master shall bring him unto the judges; he shall* *also bring him to the door, or unto the door post; and his master shall bore his ear through with an awl; and he shall serve him forever.* (Exod. 21:5, 6)

The Bible tells us that the marital bond between a man and his wife should be inseparable. His wife is to be so much a part of him that he can no longer see himself separately from her. A wife should trust, honor, and respect her husband as God's representative.

*Nevertheless let every one of you in particular so **love his wife** even as himself; and the wife see that she reverence her husband.* (Eph. 5:33)

Love toward friends:

I'm reminded of the friendship that David shared with Jonathan. Their friendship that was so close that they even made a covenant that would remain until death.

*Then said **Jonathan** unto David, Whatsoever thy soul desireth, I will even do it for thee.* (1 Sam. 20:4)

*And it came to pass, when he had made an end of speaking unto Saul, that the soul of **Jonathan** was knit with the soul of David, and **Jonathan** loved him as his own soul.* (1 Sam. 18:1)

*Then **Jonathan** and David made a covenant, because he loved him as his own soul.* (1 Sam. 18:3)

Love toward our enemies:

We are instructed by the Lord to love our enemies—not rendering evil for evil, but overcoming evil with good. Love not the evil deeds that they do. But, learn to appreciate your enemies as instrumental chastening tools to help correct and redirect you towards the Heavenly Father.

*But I say unto you, **Love your enemies**, bless them that curse you, do good to them that hate you, and pray for them which despitefully use you, and persecute you;* (Matt. 5:44)

*But **love** ye **your enemies**, and do good, and lend, hoping for nothing again; and **your** reward shall be great, and ye shall be the children of the Highest: for he is kind unto the unthankful and to the evil.* (Luke 6:35)

You may wonder why so much covering of the topic of love . . . Well, that's because without the enabling effect of God's love, there would be nothing as we know it . . . Because of God's Love, we do exist. Love is the framework and the principle by which all things exist and continue. All of our works would be useless and ineffective without it. Even our spiritual gifts operate by way of the active atonement of God's Love and Power, and moves through us to give. Being a recipient of His Love, we are to love in the same fashion, if we are born-again believers in Christ.

The Lord holds our love with high regard, which He desires to be shared toward each other in fellowship. If you're truly born again, and in a blood covenant with God the Father, then there ought to be signs of a close loving relationship that bears evidence of the lingering presence of God's love. Like the anointing oil that leaves a sweet residue of atonement. We are made to be a blessing.

*1 John 4:7 **God Is Love** [1 John 2:7] Beloved, let us [1 John 3:11] **love** one another, for **love is** from **God**; and [1 John 5:1] everyone who **loves is** [1 John 2:29] born of **God** and [1 Cor. 8:3; 1 John 2:3] knows **God**.*

The word tells us that being able to debate about doctrinal differences may make one seem important, however, our motivation should be one of love. The source of our power as the Church is in the love of God.

*[Food Sacrificed to Idols] Now regarding your question about food that has been offered to idols. Yes, we know that "we all have knowledge" about this issue. But while knowledge makes us feel important, it is **love** that strengthens the church.*
(1 Cor. 8:1)

God really recognizes and gives special attention to those who express their love to Him. Scripture tells us that He dwells in the midst of praise. Worshipping the Lord acknowledges who He is. Prayer brings about an understanding as to why there's a Lord over our lives. We should seek out opportunities to show our adoration to Him in our worship, and express our thankfulness for all that He's done by our due praises to Him for His abundant blessings, mercies, and grace.

*But the person who **loves** God is the one whom God recognizes. [Some manuscripts read the person who **loves** has full knowledge.]* (1 Cor. 8:3)

No matter how many different spiritual languages are spoken, how beautiful they may sound, or even how articulately they're spoken out of your mouth, unless the words are being spoken in love, they are fruitless. Tongues spoken out of order, and without love, are just a lot of noise before God. They are irritating to His ear.

> [**Love** Is the Greatest] *If I could speak all the languages of earth and of angels, but didn't **love** others, I would only be a noisy gong or a clanging cymbal.* (1 Cor. 13:1)

A great spiritual gift such as prophecy, or the ability to understand the deep secrets of God, or having the knowledge and the faith that causes great manifestations—without godly love toward others, would be counted as useless before God. In other words, God is not glorified through these deeds unless they are being yielded in godly love, by faith and through the Holy Spirit . . .

> *If I had the gift of prophecy, and if I understood all of God's secret plans and possessed all knowledge, and if I had such faith that I could move mountains, but didn't **love** others, I would be nothing.* (1 Cor. 13:2)

> *If I gave everything I have to the poor and even sacrificed my body, I could boast about it; [Some manuscripts read sacrificed my body to be burned.] but if I didn't **love** others, I would have gained nothing.* (1 Cor. 13:3)

Bless God! Now isn't this an eye-opener? Beloved, no matter how great the sacrifice is . . . even to the point of becoming a martyr, or giving your own body to death. If these things are being done by some other motivation beyond the sake of showing God's love toward others, then this is considered a self-glorifying act. In other words, the motives are off because it's done to bring attention to oneself, rather than God. And know that God will not share His glory with anyone. The flesh cannot be glorified. No flesh will glory in His sight!

Love *is patient and kind.* **Love** *is not jealous or boastful or proud* (1 Cor. 13:4)

Praise the Lord! Godly love enables endurance towards others. We are to be long-suffering, knowing the frailty of humanistic abilities. Being empathetic to consider your own faults, while remaining vigilant against weakness by ministering through the power of reconciliation in love . . .

This love is not out to control, nor possess what it should not. It's not about selfishly regarding oneself. However, we should love ourselves without being self-centered. Instead, the value is about God, and sharing His love toward others freely. As we have received God's love freely, it must also to be given freely toward others. Again, this love is not exhibited through a prideful spirit. You can obtain a proper spiritual perspective by checking and evaluating the true source of your motives to see if you're serving in love, or directing attention to yourself.

***Love** never gives up, never loses faith, is always hopeful, and endures through every circumstance.* (1 Cor. 13:7)

Amen, godly love will never give up on you. It always believes toward the benefit of others, and is always hopeful for the best. How do we hold the position of believing? By praying for the sake of others to a God that can do anything but fail. Godly love will cause you to pray for others, and for the benefit of their well-being. Trust in the Lord, and anticipate Him to change things for the best.

*Prophecy and speaking in unknown languages [Or in tongues.] and special knowledge will become useless. But **love** will last forever!* (1 Cor. 13:8)

There will be times when these gifts will not benefit others within their hearing of us. But if we exhibit God's love toward others, it will always be useful in reaching others and drawing them in receiving the loving presence of God through faith. And this is not to shun the practice of prophecy and speaking in unknown tongues, because these gifts are still useful in ministry towards building ourselves, and others in faith. Also, when we pray in our heavenly prayer language of spiritual tongues, we are speaking mysteries, while being enlightened ourselves concerning the deep things of God. This is all while remaining in communication with God.

*Three things will last forever—faith, hope, and **love**—and the greatest of these is **love**.* (1 Cor. 13:13)

Three things shall always remain with us. The one that stands out among the others is love. Because God is the source from which all things do exist. Without God, there's nothing, and no need for faith or hope. We hope because of what God will bring to pass. We have faith due to the substance and evidence of what God has already done . . . There's love because of who God is. For He is Love!

This Is How the Church Body Should Work as One

*There is neither Jew nor Greek, there is neither bond nor free, there is neither male nor female: for ye are all one in **Christ Jesus**.* (Gal. 3:28)

We are all one in Christ! No matter what your nationality, gender, economic status, social background, family lineage, physical genetics, or previous religious persuasion. This includes you or your family before salvation. At the point of believing the gospel, and becoming born again, each believer becomes one in Christ. Becoming joint heirs through adoption into the Family of God. Spiritual identity changes are all part of becoming new creatures in Christ. Believers are to put off the old life, and put on the new Life in Christ. Having a renewed mind that is transformed into the Mind of Christ. Therefore, the redeemed of the Lord are no longer separated from God, but are instead joint heirs with Him.

Unification through Christ allows no room for division, class separation, partiality, or discrimination within the Body of Christ. Although, we are many members possessing various gifts, talents, and functions, we are all fitly joined together into one spiritually governed Church Body.

You may wonder why I include so much scripture. Let there be no question that what I am writing comes from the word of God. I am not rewriting the Bible, but only attempting to teach and bring more clarity to the readers concerning various biblical precepts, concepts, and applications of truth.

Ephesians 4

Verse 1: I therefore, the prisoner of the Lord, beseech you that ye walk worthy of the vocation wherewith ye are called

We are charged to stay true to our own particular calling and purpose to exercise the gift, while also making good use of our talents to the best of our abilities.

Verse 2: With all lowliness and meekness, with longsuffering, forbearing one another in love;

Also, we are to be humble and submissive toward one another with patience, considering each other's burden, and then offering assistance in love.

Verse3: Endeavoring to keep the unity of the Spirit in the bond of peace.

Make an effort to promote and encourage unity among believers in the Spirit and Mind of Christ, while remaining in the fellowship of God in peace.

Verse 4: There is one body, and one Spirit, even as ye are called in one hope of your calling;

Know this: There is one spiritual Christ Body and one Spirit. Our inspiration and purpose for why we are designed to do what we do comes from the One God who gives according to His own council and desire. There shouldn't be any private agendas among believers because we all must work as one in Christ.

Verse 5: One Lord, one faith, one baptism

Believers are to serve the Lord who is the ruler and maker of all. Have no other faith outside of Him as Lord and God. All must come through one baptism into Christ, by His Name. Let's be a witness by our faith, the life that's lived, and by way of service in the beauty of holiness.

Verse 6: One God and Father of all, who is above all, and through all, and in you all.

There is only One true God and Father of all mankind, who is supreme over all, and who works and supplies all that exists. He's the source of all that is. Nothing is outside of Him!

Verse 7: But unto every one of us is given grace according to the measure of the gift of Christ.

Every one of us has been given a certain portion of grace according to the level of the gift of Christ. The Lord gives us all that we need in order to present the gift to be a blessing.

Verse 8: Wherefore he saith, When he ascended up on high, he led captivity captive, and gave gifts unto men.

Jesus Christ made it possible for us to receive spiritual gifts by the work that was performed on Calvary's cross through His crucifixion, and the resurrection of His Body that ascended up into heaven. Therefore, canceling Satan's worldly authority and restoring our authority to take dominion again over the earth, while pouring gifts upon us through the sending of the Holy Ghost.

Verse 9: Now that he ascended, what is it but that he also descended first into the lower parts of the earth?

Amen, The Messiah did His work thoroughly and left no stone unturned. No question, He settled the case. It was finished!

Verse 10: He that descended is the same also that ascended up far above all heavens, that he might fill all things.

Christ Jesus completed the work of Adam by presenting a better Man. He's the first fruit for all who would come through Him. Those who are chosen by God, will find trust and believe in Him by faith.

Verse 11: And he gave some, apostles; and some, prophets; and some, evangelists; and some, pastors and teachers;

God has chosen vassals that are raised up for our instructions, admonition, and correction for the establishment of His Order. God's word will assist in nurturing and maturing us in the Body of Christ.

Verse 12: For the perfecting of the saints, for the work of the ministry, for the edifying of the body of Christ:

Preachers and teachers are given the purpose of instilling within us the word of God, that we may know what is true, and do His will. The Lord calls us to be a unified body of believers for His glory. A people divided against themselves cannot stand. It is imperative that we work together as one. Come together with an understanding by adhering to godly percepts, principles, values and standards through His Word.

Verse 13: Till we all come in the unity of the faith, and of the knowledge of the Son of God, unto a perfect man, unto the measure of the stature of the fulness of Christ:

Unity builds up the Body of Christ by being on one accord. Having the same Christlike mind-set, and being made steadfast and unwavering in the truth. Being united in the faith; having the same mind set in Christ. Believers can be enlightened by the word of God through the Holy Spirit, and then grow together in the fullest capacity, and flourish in the things of God.

Verse 14: That we henceforth be no more children, tossed to and fro, and carried about with every wind of doctrine, by the sleight of men, and cunning craftiness, whereby they lie in wait to deceive;

Upon reaching full maturity in our knowledge, we should remain true to the Apostles' doctrine by remaining confident in the faith, and withstanding the deceitful influences or negative operations.

Verse 15: But speaking the truth in love, may grow up into him in all things, which is the head, even Christ:

Although believers have the truth, we should speak it in love. Not being arrogant or prideful with our knowledge of the truth, because such pride only stunts our growth, and hinders our ability to operate to the fullness of our potential in Christ. Know that pride eventually brings failure. That which matriculates from Christ, should also flow down through the words that we speak.

Verse 16: From whom the whole body fitly joined together and compacted by that which every joint supplieth, according to the effectual working in the measure of every part, maketh increase of the body unto the edifying of itself in love.

Members within the Body of Christ are closely fitted together, and held by every joint that's supplied by the powerful working of the Holy Spirit. Each member of every part having a specific measure towards the overall growth. Making increase of the Body unto the building up of itself in love.

Verse 17: This I say therefore, and testify in the Lord, that ye henceforth walk not as other Gentiles walk, in the vanity of their mind

There ought to be a difference in the way that we live our lives as Christians. We should not reflect the world's standards, their belief system, or lack of regard for God's word. The world's views and practices should not change or influence God's people, instead, believers should influence and impact the world. This is for the sake of those who are yet to be saved.

Verse 18: Having the understanding darkened, being alienated from the life of God through the ignorance that is in them, because of the blindness of their heart:

Knowing the truth, God's people should not follow after those who are jaded in their understanding. Because the unsaved are separated from God. Instead, they're being lead by their own stubbornness and deceitful heart. The ungodly are unable to see or comprehend godly truth.

Verse 19: Who being past feeling have given themselves over unto lasciviousness, to work all uncleanness with greediness.

Besides, the unsaved are insensitive to the fact that they are in the wrong, because they haven't any pure consciousness about themselves. Basically, their sinful desire to practice that which is unholy, is due to an uncontrollable desire to gratify physical urges and greediness.

Verse 20: But ye have not so learned Christ;

If you habitually behave as the doubtful and unbelieving, it is because you have not known Christ as Savior.

Verse 21: *If so be that ye have heard him, and have been taught by him, as the truth is in Jesus:*

If you have heard the Gospel and have been taught by His Word, then you should know the truth, because Jesus is the truth and the light. In Him is no darkness at all.

Verse 22: *That ye put off concerning the former conversation the old man, which is corrupt according to the deceitful lusts;*

By knowing the truth, you are made free! And you should put off the old ungodly nature that used to dominate you in your former life as an unsaved person.

Verse 23: *And be renewed in the spirit of your mind;*

Change and transformation comes by way of the Holy Spirit. This refreshes the spirit of man, and renews the mind daily to reach a level of operation within a transformed mind in Christ.

Verse 24: *And that ye put on the new man, which after God is created in righteousness and true holiness.*

Christians should put on the spiritually sanctified nature that was created by God. That which is made after His own image, and is in right standing with Him in true holiness.

Verse proximity. There should be **25:** *Wherefore putting away lying, speak every man truth with his neighbour: for we are members one of another.*

Since believers are positioned and fashioned in righteousness and truth, then we should stop lying to each other. Instead, speak the truth to those who are in close proximity, in fellowship, or in any sort of relationship with you. Be mindful to hear one another and speak truthfully, while sincerely empathizing and regarding them as a part of yourself, having been united in the Body of Christ.

Verse 26: *Be ye angry, and sin not: let not the sun go down upon your wrath:*

Uncontrollable anger is a deadly and destructible tool of the enemy. Festering anger can do damage to both the carrier and the receiver of it. Carriers who are instruments of anger are straying outside of God's true fellowship. Carrying such anger would subject this rebellious vassal to suffer other oppressive encounters with the enemy. Anger that is left unchecked or overlooked, will grow and develop into another area, like a poison or cancer. Its only impulse is to destroy. Anger does not discriminate or even respect the one who's harboring it. In fact, if not done away with or rejected, it will find its way into the heart of the person bearing it. Once inside, it will send signals through out the physical body as well as the spirit of the individual. Its mission is to seek, capture, and destroy by attacking the physical body with stress-related disease. Moreover, anger also works to limit spiritual awareness and hinder the spiritual relationship one may have with the Heavenly Father. The remedy is to repent, have a spiritual detox; by way of receiving the wholistic healing value of God's Word, through the Power of the Holy Ghost. The Blood of Jesus purifies us of impurities, releases the Holy Ghost, while giving us new life and power to overcome all trials. The Holy Ghost also enables us abilities to do great exploits to the glory of God.

Verse 27: Neither give place to the devil.

The enemy tries to infiltrate the Body of Believers through combative approaches such as: division, stealing, deception, manipulation, destruction, and death. This spiritual wickedness come in the form of evil divisive subcommunities and colonies. Let's not give the devil easy access to work within our midst through devices that work to divide the Body of Christ.

Verse 28: Let him that stole steal no more: but rather let him labour, working with his hands the thing which is good, that he may have to give to him that needeth.

Those who stole in the past, should not steal anymore, but endeavor to work with your hands. To do that which is good and prosperous, so that you may have to give to those who are in need.

Verse 29: Let no corrupt communication proceed out of your mouth, but that which is good to the use of edifying, that it may minister grace unto the hearers.

Don't allow evil or negative words to be spoken out of your mouth, but speak what is good for strengthening and building up, so that it may minister grace to the hearers.

Verse 30: And grieve not the holy Spirit of God, whereby ye are sealed unto the day of redemption.

Don't mourn the Holy Spirit of God, because His love is shed abroad over our hearts, and He has put His seal of approval upon us. The Holy Spirit works to keep us until the Lord comes back for us.

Verse 31: Let all bitterness, and wrath, and anger, and clamour, and evil speaking, be put away from you, with all malice:

Put away bitterness, fits of rage, anger, chaos, evil speaking, negative attitudes, and actions that constitutes malicious behavior.

Verse 32: And be ye kind one to another, tenderhearted, forgiving one another, even as God for Christ's sake hath forgiven you.

Instead, be kind to one another and compassionate, forgiving one another, even as God for Christ's sake has forgiven you.

How Far Shall We Believe What the Pastor Says?

Pastors should be accountable to God, their spouse, children, and their congregation. In that order, with God being first and foremost. The reason being, is that God is to be revered and honored above all. And if the Pastor's accountability is to God first, then all others will reap the benefits of a chastened vassal that has an excellent spirit of tried integrity. This individual is capable of loving and showing others appropriate consideration, acknowledgement, and respect. A sign of a good leader is marked by the standards that are kept within his or her own household. This is not to say that there won't be problems, but through it all, the person is able to maintain a loving environment of commitment and cohesion within the household.

The Bible tells us that faith comes by hearing. How can we hear except by a preacher? By God's design, a preacher is a necessary component of our faith-building process. By faith, we are to believe that God has placed us in, or is directing us to a safe haven to be fed and nurtured in the things of the Lord. For believers, our safe haven is found in the gathering of corporate Christian believers. We are admonished not to forsake our assembling together for corporate worship, prayer, fellowship, or from listening to the preached word of God. Even more so, as the day approaches, and signs of the times indicate that the coming of the Lord is near. There's safety in the midst of the flock that's near its shepherd. The Lord, being our Shepherd, provides us with a shepherd that is above all other shepherds. He's the Good Shepherd, who has proven that He will lay down His life for us. Our Lord will feed us when we're hungry, fight off our enemies, make us lie down when we are weary, restore us with strength, correct us, and lead us in the right path. There's a prepared place of refuge and covering in the camp of our God.

We must trust the Lord to place us under true leadership. By faith, accept the minister of God as the one sent to speak words of life that should produce spiritual maturity. But nevertheless, the Bible has warned us that there are many false prophets out to deceive many. How do you recognize a true prophet, pastor, or teacher? We can't always judge what's in a pastor's heart, but you can hold them accountable to stick with the scriptures. We are called to submit and obey as God leads us to. If it's not scriptural, then you're not required to follow or submit to it. However, I must also say that no pastor is 100 percent right in everything he or she talks about, or does. Only God's absolute truth, is constant and never varies. But because of the frailties of man, there is room for a minimal percentage of human error. I did say minimal room, but if the person turns out to consistently stray or veer off from what is biblically true, then that person is not capable of teaching the word of God. And consequently, should not occupy a position to instruct a congregation. In such case, you are warned to leave, prayerfully seeking God's direction towards another pastor to encourage and equip you in the ministry of the Lord. But don't stay out of church fellowship! Get planted quickly into a more biblically sound ministry, and fellowship with true believers. The Bible instructs us to follow religious leaders as they follow Christ. In other words, as they follow after the things of God, according to His divine ordinates. Congregates are to be actively involved by combining their own individual studies with ongoing participation in church studies to better learn the Bible for what it truly says.

Study to show thyself approved unto God, a workman that needeth not to be ashamed, rightly dividing the word of truth. (2 Tim. 2:15)

How Should We React to Spiritual Demonstrations?

Believers are called the peculiar people of a royal priesthood that is under divine headship. This puts us in a position as priestly people who follow after Christ. How should we react to spiritual demonstrations within a church? Well, when it comes to spiritual things such as laying of hands. The act of laying hands would be prompted through prayer to touch with the hands as a point of contact. Laying of hands would be for the purposes of: impartation for receiving the Holy Spirit, healing the sick, prayer for deliverance, placing a blessing, acknowledging a gifting, and appointing authority. In addition, this was practiced in the Old Testament. Levitic priests performed this practice for sin atonement. They would lay their hand upon the head of a sacrificial animal before killing it, or sending it out into the wilderness. The priests would carry out this ceremonial act of laying hands on an animal for the purposes of: transference of sin before putting the animal to death, or reversing a curse onto the animal. This was also done prior to placing judgment on an individual before execution. Those old sin atonement practices of the use of animals are no longer acceptable in the sight of God, because He provided the Eternal Sacrificial Lamb through Christ Jesus for our sins. Since the birth of the Church, there are other reasons for laying on of hands. Christians do it to bless and empower each other for the work of the Kingdom. However, the enemy comes to deceive many to serve another purpose towards ungodly actions. So beware, there can be unrighteous reasons behind the action. It depends on who's laying the hands, and for what purpose. Mind you, discern whether they are sent by God to bless you. I take it that if you are already being blessed by the one who is preaching the word, then you can be relieved that blessings of impartations, signs, and wonders will follow as well. When the Gospel is preached, there should be demonstrations of power that follow through the word as it is received by faith. Lives ought to be changed; souls ought to be saved!

This is a gift that should be practiced and not abandoned.

Neglect not the gift that is in thee, which was given thee by prophecy, with the laying on of the hands of the presbytery. (1 Tim. 4:14)

Laying hands to heal the sick:

And besought him greatly, saying, My little daughter lieth at the point of death: I pray thee, come and lay thy hands on her, that she may be healed; and she shall live. (Mark 5:23)

For healing:

And it came to pass, that the father of Publius lay sick of a fever and of a bloody flux: to whom Paul entered in, and prayed, and laid his hands on him, and healed him. (Acts 28:8)

Signs of those who believe:

They shall take up serpents; and if they drink any deadly thing, it shall not hurt them; they shall lay hands on the sick, and they shall recover. (Mark 16:18)

You cannot purchase this gift, and should be mindful as to the true motivations for receiving it.

Saying, Give me also this power, that on whomsoever I lay hands, he may receive the Holy Ghost. (Acts 8:19)

A deceitful heart cannot receive the Holy Ghost.

And when Simon saw that through laying on of the apostles' hands the Holy Ghost was given, he offered them money (Acts 8:18)

Receiving impartations of gifts and callings:

*Neglect not the gift that is in thee, which was given thee by prophecy, with the laying on of the **hands** of the presbytery.* (1 Tim. 4:14)

Be mindful or careful not to suddenly lay hands on others. And neither should you allow someone's hands to be lain casually upon yourself, without knowing or discerning the motive behind it. Depending upon who is laying hands; they can either impart a blessing or an unclean or oppressive impartation.

***Lay hands** suddenly on no man, neither be partaker of other men's sins: keep thyself pure.* (1 Tim. 5:22)

Righteous Conduct In the Church Body of Christ

Romans 12

Verse 1: I beseech you therefore, brethren, by the mercies of God, that ye present your bodies a living sacrifice, holy, acceptable unto God, which is your reasonable service.

   This is an earnest plea for those who are called into the fellowship of believers to dedicate and submit our whole being sacrificially, in service to God.

Verse 2: And be not conformed to this world: but be ye transformed by the renewing of your mind, that ye may prove what is that good, and acceptable, and perfect, will of God.

   We are not to allow the ways of the world to dictate the way that we should live our lives, but should rather set godly examples for those who would see the light, and then learn from it.

Verse 3: For I say, through the grace given unto me, to every man that is among you, not to think of himself more highly than he ought to think; but to think soberly, according as God hath dealt to every man the measure of faith.

Verse 4: For as we have many members in one body, and all members have not the same office:

Believers are a synergy of many joined members in one Spiritual Church Body, and all members occupy various positions, functions, and expressions of gifts and purposes.

Verse 5: So we, being many, are one body in Christ, and every one member one of another.

And although we are many members, yet we are one Body in Christ, and every member is connected to each other.

Verse 6: Having then gifts differing according to the grace that is given to us, whether prophecy, let us prophesy according to the proportion of faith;

To whom much is given, much is required. Each member has different gifts according to the level of grace that's proportionally activated by the measure of faith that's given to us. If prophecy is our gift, then let us prophesy according to the proportion of faith that works within us. We should work together, perfecting and utilizing our gifts as we are anointed to do so. Let all members remain within their own calling, and at the level by which God ordains it to be. Stay true to your calling. Don't covet another's gift, or try to imitate or mimic another's gift. In other words, don't be an imitator, just be the original that God created you to be. No one can be better at being you than yourself. All of God's creations are original and unique. Something that seems simple to our eyes beholds unique differences of information that stems far beyond our understanding. Consider the common snowflake, it is yet uniquely designed and masterfully thought out so that no one snowflake bears the same identical pattern of design. Wow! What an awesome God is He! Imagine if we can witness the Glory of God through something as simple as a snowflake. How much more can we, who are fashioned in His image, give Him Glory by simply being all that He called and intended for us to be.

Verse 7: Or ministry, let us wait on our ministering: or he that teacheth, on teaching;

New converts and babes in Christ should not rush into a position of ministry. Instead, you should wait to receive God's call, instructions, and preparation through a time of hearing from the Lord. Receive understanding, and then become more knowledgeable of the word of God. Learn by enduring testing and trials of your faith, while growing in maturity that proves evidence of your stability in the truth. Not everyone is called to preach or teach. Instead, wait for your gift to surface and become identifiable, whereby you are able to shine and give God the glory. The gift is not for your personal gratification. It is given to you, so that you may glorify God through the gift. The glory goes to God and God alone. He will not share His Glory with another, and neither will any flesh glory in His sight Isaiah 48:10-11.

Verse 8: Or he that exhorteth, on exhortation: he that giveth, let him do it with simplicity; he that ruleth, with diligence; he that showeth mercy, with cheerfulness.

The ability to exhort requires a certain level of experience and maturity in the Lord. This gift and all other ministerial offices must be accompanied and exercised by inspiration of the Holy Spirit. In order to effectively urge others to a higher level of godly conduct, there must be the presence of the Holy Ghost to minister though the chosen vassal to speak words that impact and promote change within the heart of the hearers.

Having the office of exhortation would entail encouraging others towards commendable conduct. The objective would be to: urge the call to repentance, continue in the faith, convict gainsayers, warn the unruly, encourage soberness, strengthen godliness, and to stir up liberality of godly truth. There are special times when it is necessary to encourage others to adhere to sound doctrine (2 Tim. 4:2-5).

Now, if you are called to be a giver, do so without being overly conspicuous. Just simply do it in the name of the Lord, not requiring any praise to yourself, or repayment. If you are called to have authority over others, administer your authority with carefulness and consideration for those who you have been placed over. Those who are called to show mercy, should do so with a cheerful heart and tenderness.

Verse 9: Let love be without dissimulation. Abhor that which is evil; cleave to that which is good.

Let's present love without deceit or variance. Let's not seek opportunities to manipulate others for the purpose of satisfying selfish desires that lead to unprofitable gain. Be true to love. Don't act as though you love, and on the other hand act as if you don't. Resist evil by denying any opportunity to yield to it. This is done by avoiding contact with anything that even references evil. Believers are to even stay away from the very appearance of doing evil. Only embrace that which is good. God is Good!

Verse 10: Be kindly affectioned one to another with brotherly love; in honour preferring one another;

We are to treat each other with loving kindness, respectfully interacting with each other in a spirit of humility.

Verse 11: Not slothful in business; fervent in spirit; serving the Lord;

Don't be slow or lazy in handling business matters. But rather rather be zealous, with an understanding and diligence to your course in service to the Lord.

Verse 12: Rejoicing in hope; patient in tribulation; continuing instant in prayer;

Pray to the Lord and rejoice in the hope of the God of our salvation. Be thankful unto Him in all situations. Trust and have faith in the Lord, as you patiently witness His Hand of deliverance. His protection and guidance will lead you out of every trial and tribulation with Power that brings Glory to Himself. Develop such a relationship with the Lord that you maintain continuous and open communication with Him. In all of our ways we are to acknowledge Him, and He will direct our path.

Verse 13: Distributing to the necessity of saints; given to hospitality.

We as the Church should distribute and exchange whatever resources, information, knowledge, and ideas that's available among each and every one of us. This works to meet needs within the Body of Christ, while also enabling the Body to contribute to each other without bias. If you are privileged to be in a position of abundance, and can share freely, then do so. Be a blessing to those who are less fortunate. Don't withhold something that could benefit someone who needs it. Try not to withhold services, materials, money, knowledge, or information for the sake of appearing better than you really are, or for competitive reasons. You won't prosper by doing that. Be generous and kind toward everyone. Welcome fellow believers, be communal, and extend arms of hospitality through fellowship, socializing, sharing, and enjoying one another's companionship. In doing so, we find opportunities to encourage and strengthen fellow believers, while welcoming new converts to Christ. This benefits the Body of Christ toward the advancement of the Kingdom of God.

Verse 14: Bless them which persecute you: bless, and curse not.

Be a blessing even to those who strive to work against you. Don't speak evil or desire their punishment. Hold no unforgiveness or bitterness toward them. Allow yourself to exhibit Christ-likeness that your enemies can see. In doing so, they will be astounded and amazed at the level of strength and excellent character that God has bestowed upon you. Choose to uphold good. Many come to Christ by this wonderful display of love in action.

Verse 15: Rejoice with them that do rejoice, and weep with them that weep.

As fellow believers in the Lord, you are to be empathetic, sympathetic, and moved to respond to one another's burdens or grief. We're not to carry any feelings of jealousy, envy, bitterness, or hatred. Instead, we should relate to other born-again Christian believers as one in the same, being sensitive to each other's feelings of joy and pain. You must be able to rejoice with those who are experiencing joy, and be glad for them as though it was yourself experiencing the blessing!

Verse 16: Be of the same mind one toward another. Mind not high things, but condescend to men of low estate. Be not wise in your own conceits.

When it comes to the things of Christ and fellow believers, be able to come together in the fellowship of agreement with the Holy Spirit. Have no private agendas of your own, and do not think of yourself as being better than anyone else. In doing that, you would only fool yourself in thinking that you could ever be a true success apart from having godly fellowship. Otherwise, put on the Mind of Christ, and be transformed by the renewing of your mind. Allow your mind to be washed and regenerated through the Word of God. In doing so, all are equal and unified as one in Christ, and are then able to occupy various positions or fulfill various missions for the common goal of the Kingdom in Christ.

Verse 17: Recompense to no man evil for evil. Provide things honest in the sight of all men.

Believers are not to repay anyone evil for evil. It is certainly not our position to act as an evil judge, or as the instrument of a foul, cantankerous, or obnoxious spirit. As an alternative, we are to uphold a certain level of integrity in the sight of others. Our living example is under heavy scrutiny by those who are just waiting for an opportunity to justify their sinful nature and unbelief towards God. The world does not want to believe that there is a God who sits High, as the Righteous Judge. Let's not make it easy for them to remain in a state of unbelief!

Verse 18: If it be possible, as much as lieth in you, live peaceably with all men.

If it is at all possible, try as hard as you can to live in peace. Furthermore, strive to keep peace with everyone that you encounter in fellowship, or reside with. Now, if you have to act in a defense mode as God leads, then move and respond as the Holy Spirit instructs you to do. There is a time for peace and a time for war. However, let not the thought of vengeance or war enter your minds to act out, or speak out against other Christians. But trust the Lord, and adhere to His voice of reason, and never blame God for of any evil doings or folly. Rejoice in everything, and give thanks to God, knowing that there is a time for everything.

Verse 19: Dearly beloved, avenge not yourselves, but rather give place unto wrath: for it is written, Vengeance is mine; I will repay, saith the Lord.

Don't repay evil for evil! Vindictive activities and responses should not be a part of our personality traits. We have to forgive, and not hold grudges toward others. You can be angry without sinning. It's when you move from the point of anger to act out vindictively, that it becomes an act of sin. However, there is such a thing as having a holy indignation or righteous anger against that which is evil. We are encouraged to hate that which God hates, and to love what He loves. God hates all acts of sin. So, we can be angry without sinning, as long as we are prompted to move or respond in a righteous manner. Besides that, it is not for us to retaliate against those who have hurt or harmed us. We must rely on the Lord to intercede on our behalf, and righteously assess the situation, then provide remedy for judgment and repayment. The Lord is our Protector from all hurt, harm, and danger. He gives us the strength when it's time to fight. Our God is a God of perfection, and He will correct whatever needs to be corrected in its due time. He leaves nothing undone, sees everything, knows all things, and makes no mistakes!

Verse 20: Therefore if thine enemy hunger, feed him; if he thirst, give him drink: for in so doing thou shalt heap coals of fire on his head.

We don't have to return evil with evil. Instead, we can win over our enemies by our capacity to show compassion, love, and caring. Our contrast to their evil actions will offer a mirrored reflection of their present image, showing them the evil that they posses. Through this method, an irritant is produced that could serve as an element of shame towards their immoral behavior.

Verse 21: Be not overcome of evil, but overcome evil with good.

Let's not allow ourselves to be consumed by the evil and degenerate influences to the point where it shapes our identity. By doing so, we take on the image, ideology, and belief systems of the world. Instead, let us suppress evil by exemplifying the Christ likeness that is within us. We then allow our light to overcome whatever darkness and evil there is in this world. Furthermore, we offer brighter solutions and better alternatives to life's circumstances through Christ. We have the keys to eternal life through the Gospel of Jesus Christ. Let's share the keys with those who are in danger of death or destruction. And instead, permit them to witness the abundant life that we are so blessed to enjoy.

Was Nicodemus Saved?

Nicodemus asked Jesus what he should do to be saved. Jesus answered Nicodemus by saying that he must be born again. Of course, Nicodemus was being carnally minded at that time, or was thinking from a physical point of view. He could not comprehend what this spiritual statement meant, but in turn, asked how he could be born again except he enters into his mother's womb again. Sometimes, a person can intellectualize through worldly wisdom, and miss out on the gospel truth. Nicodemus missed the truth because it was hidden from him due to the spiritual condition of his heart. Jesus went on to say that unless you be born from above, that you will not enter into, comprehend, or see the Kingdom of God. Jesus told him that he must be born of the water and of the Spirit. Nicodemus was not yet ready to understand this saying. Jesus let Nicodemus to know that he must not reject the truth, but instead, believe on Him. Amen . . . Nicodemus, through being open to the truth, eventually grasped on the the truth. He later became a follower of Jesus Christ, as he was one of those who helped to receive the crucified body of Jesus. Amen. Nicodemus was saved.

John 19:39

[John 3:1] **Nicodemus**, *who had first come to Him by night, also came [Mark 16:1] bringing a mixture of [Ps. 45:8; Prov. 7:17; Song of Sol. 4:14; Matt. 2:11] myrrh and aloes, about a [John 12:3] hundred pounds weight.*

> (BL)Nicodemus, who had first come to Him by night, also came, (BM)bringing a mixture of (BN)myrrh and aloes, about a (BO)hundred pounds weight. So they took the body of Jesus and (BP)bound it in (BQ)linen wrappings with the spices, as is the burial custom of the Jews. Now in the place where He was crucified there was a garden, and in the garden a (BR)new tomb (BS)in which no one had yet been laid. Therefore because of the Jewish day of (BT)preparation, since the tomb was (BU)nearby, they laid Jesus there. (John 19:39-42)

But we are not to reject the truth by disbelief. The Pharisees were also disobedient. Furthermore, not trusting or relying on Jesus will bring condemnation instead. There is a difference between being spiritually religious as opposed to being superficially and hypocritically religious. Furthermore, being religious is not a bad thing, if faith applications includes the Will, Purpose, and Power of God . . .

Mark 1:10

*And when He came up out of the **water**, at once he [John] saw the heavens torn open and the [Holy] **Spirit** like a dove coming down [[Kenneth Wuest, Word Studies.] to enter] [Literal translation of eis.] into Him. [John 1:32]*

These are the two symbolic witnesses upon the earth that testify of Jesus Christ.

*John answered them all by saying, I baptize you with **water**; but He Who is mightier than I is coming, the strap of Whose sandals I am not fit to unfasten. He will baptize you with the Holy **Spirit** and with fire. (Luke 3:16)*

John, who was a forerunner, declared of Jesus the Anointed One, was coming with an even greater work. And that Jesus Christ is far better than himself; whose sandals he was not worthy to unloose. John saw himself as an unworthy servant. John said that he must decrease and the Lord's Will, Purpose, and Plan must increase.

*For John baptized with **water**, but not many days from now you shall be baptized with ([Kenneth Wuest, Word Studies in the Greek New Testament.] placed in, introduced into) the Holy **Spirit**. (Acts 1:5)*

This scripture revealed that although John baptized the unrighteous person unto repentance, there was another baptism. Also, that the baptism of the Holy Spirit would baptize the believer with fire beginning at the day of Pentecost.

*This is He Who came by (with) **water** and blood [[Marvin Vincent, Word Studies.] His baptism and His death], Jesus Christ (the Messiah)—not by (in) the **water** only, but by (in) the **water** and the blood. And it is the [Holy] **Spirit** Who bears witness, because the [Holy] **Spirit** is the Truth.* (1 John 5:6)

The Holy Spirit bears witness internally and externally, that we are the children of the Most High God. The Spirit bears witness of Jesus Christ and the manifested life in Christ. There is also the witness of His water baptism and His blood purchase through death. The Holy Spirit being the witness in truth concerning the two; being altogether three witnesses which bear witness of each other. Let's read 1 John 5:8: *"And there are three witnesses on the earth: the **Spirit**, the **water**, and the blood; and these three agree [are in unison; their testimony coincides]."*

Revelation 22:17

*The [Holy] **Spirit** and the bride (the church, the true Christians) say, Come! And let him who is listening say, Come! And let everyone come who is thirsty [who is painfully conscious of his need [Joseph Thayer's Greek-English Lexicon] of those things by which the soul is refreshed, supported, and strengthened]; and whoever [earnestly] desires to do it, let him come, take, appropriate, and drink the **water** of Life without cost. [Isa. 55:1]*

This tells us that the Holy Spirit cannot be purchased, but instead, must be earnestly desired and thirsted after like a deer thirsts after a watery brook. Knowing that for our survival we need this infilling to be complete.

Let's cover the subject of "speaking in tongues." Is it really a necessity? Well, the answer is Yes! Since Jesus said that it would be one of the signs of those who believe. I'll back this statement up with some scripture. Turn your Bibles to Mark 16:15 and conclude at verse 20; it reads . . .

Verse 15: "And he said to them, Having gone to all the world, proclaim the good news to all the creation;"

Verse 16: "He who hath believed, and hath been baptized, shall be saved; and he who hath not believed, shall be condemned."

Verse 17: "And signs shall accompany those believing these things; in my name demons they shall cast out; with new tongues they shall speak;"

Verse 18: "Serpents they shall take up; and if any deadly thing they may drink, it shall not hurt them; on the ailing they shall lay hands, and they shall be well."

Verse 19: "The Lord, then, indeed, after speaking to them, was received up to the heaven, and sat on the right hand of God;"

Verse 20: "And they, having gone forth, did preach everywhere, the Lord working with [them], and confirming the word, through the signs following. Amen."

Now, let's absorb some of the context of these scriptures, starting at verse 17, noting that these shall be the signs of those who believe; they shall cast out devils, meaning that they shall possess God-given authority. It is this authority, by which even demons will be subject to, and they must obey our command to leave and cease from operating within a place that is declared Holy, or is sanctioned to be out of bounds. Believers will also demonstrate the ability to with new tongues, which means they will speak and uncommon language to the unbelieving. that they will speak an uncommon language to the unbelieving . . . These tongue-talking people will have a common language among other believers, although our utterances will be different from one another. And we will have our own personal prayer language, all things will remain common amongst each other. We will be able to understand various prayer languages through the spirit of interpretation. The speaking in other tongues is listed among the signs as evidence of those who believe and are saved. For it states in verse 16 of Mark, chapter 16, that those who believe and are baptized shall be saved and those who don't, would be damned, as this scripture says . . . There are those who make it their business to explain the necessity of being baptized and receiving the Holy Ghost with evidence of speaking in tongues away. But let the Holy Bible be the final word.

Verse 18 says that we will take up serpents, and if we drink any deadly thing, it shall not hurt the believer, meaning the believer is to have power over satanic forces or any unexpected events that would try to destroy us. Moreover, we will be able to fight against any method the enemy devises to keep us from performing the Will and Purpose of God in our lives.

The same verse shows that believers will be able to lay hands on the sick, and they will be healed. This is a sign of imputed authority unto the believer to declare God's word, by coming into agreement with the redemptive work of Christ Jesus on Calvary's Cross.

By what authority do I speak these things? The Lord gave this command on the day of His resurrection, just before He ascended up into heaven (Mark 16-19), and after He had spoken this, He was received up into heaven. Jesus being the final authority, commissioned each of His disciples (verse 15) to go into the world and preach the Gospel to every creature. We're not alone in doing Kingdom work, and spreading the gospel. Mark 16:20 says that they preached everywhere, and the Lord was with them, after He had ascended to heaven and sat on the right hand of God. Henceforth, the Word went forward with signs and wonders following. Believers, and especially leaders need to just believe God. Do what the Lord says, and allow the Holy Ghost to move in the midst of the people. We can't be afraid to let God showcase his presence. The Heavenly Father wants to prove that He's real. He wants to show forth His Love toward us. Let's welcome God's Holy Spirit to abide in us and govern through us . . .

The earth was without form, and an empty waste, and darkness was upon the face of the very great deep. The Spirit of God was moving (hovering, brooding) over the face of the waters. (**Gen. 1:2**)

Before mankind existed upon the earth, God's Spirit was moving, preparing what was to exist, and setting things in place.

Marvel not! You must be born again!

John 3

Verse 1: Now There was a certain man among the Pharisees named Nicodemus, a ruler (a leader, an authority) among the Jews,

Verse 2: Who came to Jesus at night and said to Him, Rabbi, we know and are certain that You have come from God [as] a Teacher; for no one can do these signs (these wonderworks, these miracles—and produce the proofs) that You do unless God is with him.

Verse 3: Jesus answered him, I assure you, most solemnly I tell you, that unless a person is born again (anew, from above), he cannot ever see (know, be acquainted with, and experience) the kingdom of God.

Verse 4: Nicodemus said to Him, How can a man be born when he is old? Can he enter his mother's womb again and be born?

Verse 5: Jesus answered, I assure you, most solemnly I tell you, unless a man is born of water and [[a]even] the Spirit, he cannot [ever] enter the kingdom of God.[A]

Verse 6: What is born of [from] the flesh is flesh [of the physical is physical]; and what is born of the Spirit is spirit.

Verse 7: Marvel not [do not be surprised, astonished] at My telling you, You must all be born anew (from above).

What Must I Do to Be Saved?

We must not take this question lightly. The word tells us that there are two necessary witnesses that the process of salvation has taken place. And without this enisled impartation there's no way that we will enter into the Kingdom of God, and furthermore, we will not be able to comprehend or even see it. Without the born-again experience, we are lost with no hope.

Nicodemus was a leader with authority among the Jews. He came to Jesus at night to ask Him what he must do to be saved. He acknowledged Jesus as a Teacher sent from God, because he saw the signs and miracles that followed Jesus' ministry. But one might also wonder why he came at night; was he sneaking to ask the question? But anyway, Jesus answered Nicodemus by saying that he must be born again. As I stated earlier, Nicodemus missed the truth because it was hidden from him due to the spiritual condition of his heart. However, an inquiring mind is good, if you're seeking for truth. Nicodemus had such a mind that was prompted by the Lord. Jesus went on to say that unless you are born from above that you will not enter into, comprehend, or see the Kingdom of God. Jesus told him that he must be born of the water and of the Spirit. The Lord stimulates our thoughts to reflect or ponder upon His truth. We must simply trust in His Words to receive them for understanding. We are not to reject the truth by disbelief. The Pharisees were also disobedient and did not trust, or rely on Jesus. Not trusting Him would bring condemnation instead.

Also, Peter the apostle of Jesus Christ, was asked the question concerning what shall they do as brethren? You mean to tell me that they were yet being called the brethren, even though salvation was not yet fully obtained? Read Acts 2:37, 38:

Now when they heard this, they were pierced to the heart, and said to Peter and the rest of the apostles, "Brethren, (AX)what shall we do? Peter said to them, "(AY)Repent, and each of you be (AZ)baptized in the name of Jesus Christ for the forgiveness of your sins; and you will receive the gift of the Holy Spirit.

Amen, Peter told them to first repent. Repent means to have godly sorrow for the wrong you have done, recognizing that you have sinned, and do not measure up to where godly standards are appropriated. Know that you are a sinner that needs the Lord's provisions and help to be saved from death and damnation. Then call the name of Jesus to save you from a life of sin, and ultimately from eternal torment in hell. Trust in Him, and follow after His commands by submitting to His ordinances by faith. Then follow His precepts by repenting and identifying with Him through baptism in Jesus Christ's name.

You will house within your temple, either the Holy Spirit or an unclean spirit. God wants to clean and sweep up our temples to reflect the Holiness of HIM, and become suitable vassals in which HE is able to dwell within us. Christ was the Perfect Example meant to direct us towards everlasting life.

*When the unclean **spirit** has gone out of a person, it roams through **water**less places in search [of a place] of rest (release, refreshment, ease); and finding none it says, I will go back to my house from which I came* (Luke 11:24)

The indwelling of the Holy Spirit is a recognizable event. You will know it when it happens, as you experience the indwelling of His presence, and others will know God is in your life too.

*And I did not know Him nor recognize Him, but He who sent me to baptize in (with) **water** said to me, Upon Him whom you shall see the **Spirit** descend and remain, that One is He Who baptizes with the Holy **Spirit**.* (John 1:33)

We must identify with Christ by following after His way, and obeying His ordinances and dictates. Jesus Christ was baptized with water, and so must we. He received power and approval by way of the Spirit of God. Likewise, we must seek to be empowered, and receive the seal of God's approval upon our lives through receiving the Holy Spirit.

*And when Jesus was baptized, He went up at once out of the **water**; and behold, the heavens were opened, and he [John] saw the **Spirit** of God descending like a dove and alighting on Him.* (Matt. 3:16)

You must thirst after God's righteousness in order to receive the outpouring of His Spirit upon you. This is by having a desire to be filled with the Holy Ghost.

The Holy Spirit cannot be purchased, because the gift is free. This is the gift that is made available to us freely, by way of a priceless purchase exchange through Christ's death. His sacrificial Blood served as a priceless purchased gift that we could never afford.

What is the proper application for baptism?

I would again say that we are to be baptized in water by being totally submerging into water, not by the sprinkling of water, which is not correct.

There's also a spiritual baptism now taking place among believers beginning at the day of Pentecost. One can be baptized in the Holy Ghost with evidence of speaking with other tongues as the Spirit gives utterance. This shows the presence of the Holy Spirit being upon you. But you must acknowledge the Lord's presence, and welcome Him in to abide within you. He will not force Himself upon you. One must willingly accept Him in.

We have the opportunity to be baptized both physically and spiritually even today, just as Jesus told Nicodemus that we should be baptized by water and Spirit.

*I have baptized you with **water**, but He will baptize you with the Holy **Spirit**.* (Mark 1:8)

Believers identify with other believers by way of the born-again experience.

*Can anyone forbid or refuse **water** for baptizing these people, seeing that they have received the Holy **Spirit** just as we have?* (Acts 10:47)

These Shall Be the Signs of Those Who Believe

He who hath believed, and hath been baptized, shall be saved; and he who hath not believed, shall be condemned. And signs shall accompany those believing these things; in my name demons they shall cast out; with new tongues they shall speak; (Mark 16:16, 17)

There shall be the signs among those who believe. They shall cast out devils, meaning that they shall posses God-given authority. Another sign of a believer is that they will speak with new tongues. Meaning that we have the ability to speak in a exclusive language that's uncommon to the unbelieving. Tongue talking people share common languages among other believers, although our utterances will be different from one another. We will all possess our own personal prayer language, but will have all things common among each other. We will be able to understand each other through the spirit of interpretation. The speaking in other tongues is listed among the signs as evidence of those who believe, and are saved. For Mark 16:16 states, that those who believe and are baptized shall be saved and those who don't, will be damned as this scripture says . . . There are those who make it their business to explain the necessity of being baptized and receiving the Holy Ghost with evidence of speaking in tongues away. But let the Holy Bible be the final word.

God has given each believer power over satanic forces or unexpected events that would try to destroy us or prevent us from performing the will and purpose of Him in our lives. And those who believe in His Living Word will have the power to heal. We will be able to lay hands on the sick, and they shall recover. It comes with the authority of the Believer to declare, and to come into agreement of what was established on the cross by the redemptive work of Christ Jesus, the Yeshosua, meaning the Lord of Salvation in Hebrew. Here settles the work of redemption, when Christ the Savior said that it is finished!

Some Traditions Are Worth Keeping

The Church should revisit divine and universal principles of practicing ecclesiastical ethics, such as allowing some things to rest on the seventh day of each cycle to refresh. Then the Church should begin anew from a stand point of new beginnings on the eighth day. Honoring the Sabbath was meant to be a blessing, not a law of bondage. There is reflection, liberty, rest, and restoration as specific advantages (Isaiah 58:9-14). Farmers should also consider applying this principle agriculturally, by allowing the depleted soil in their crops to rest from the over usage of harvesting crops every seven years, for one year. Then return again for the soil to replenish itself, and begin a fresh harvest crop. Since we've abandoned this practice, our crops are depleted of their natural minerals and vitamins. Because, we don't allow our crop soil to rest before rotating to plant again.

Another opportunity for renewal would be in celebration of Jubilee. After every forty-nine years, Jubilee should be reinstituted as a mainstream practice for the purpose of debt reduction, price fixing, and bringing about liberty and freedom to the enslaved under conditions of bondage. Land and property owners would have their belongings restored, all of their debts would paid in full, and rest would be restored to the land.

But Thank God! This material order of things will not determine our spiritual order in Christ. Even though this world will pass away, our eternal rest and liberty is in Christ Jesus (Rom. 8:19-24).

Logic Model to Kingdom Building

Kingdom building is a supernatural act of God that employs us to be instruments of His magnificent handiwork. This is work that is not done by physical hands. God will call things in place as He sees fit. We must be spiritually aware of the clarion call of God in our lives, and be ready and available to do as the Lord desires us to do. Keep in mind that God is the Chief Builder of His Kingdom. There is no place for rebellion, pride, greed, or deceit for this work. If we have hidden agendas, transgressions, or impure motives then we will be prohibited from this reformation process. What does it profit a person to gain the world and lose their own soul . . .

On the other hand; there is evangelistic growth by earnest preaching of the gospel, and with signs of power following. Recycling members simply migrating from one congregation to another, does not legitimately constitute true church growth. Unless those migrating became born again; resulting from answering the call to salvation under the preached word of the new congregational pastor. True church growth, comes by adding new born again believers through evangelism. Also, evangelistic growth comes by awakening old and new born again believers towards revival for service.

The pastor who has a heart to minister to the unchurched; this would require a spirit like the power of Elisha. This pastor must possess a spiritual fervor for true revival and evangelism. Also the pastor must have a willingness to comment and submit to the Lord, through prayer, fasting, sanctification, in addition to walking in truth and love. A godly lifestyle costs more, it requires you to totally sell out to the Lord. Having a godly lifestyle is your salt. This salt changes the flavor of whatever it's added to; being your power to hold godly influence to the nations . . . Having power for revival comes from the Holy Ghost.

My scriptural references:

Mark 9:50 Salt is good: but if the salt have **lost** his saltness, wherewith will ye season it? Have salt in yourselves, and have peace one with another.

Matthew 16:26 For what is a man profited, if he shall gain the whole world, and **lose his own soul**? or what shall a man give in exchange for **his soul**?

Mark 8:36 For what shall it profit a man, if he shall gain the whole world, and **lose his own soul**?

2 Kings 13:21 And it came to pass, as they were burying a man, that, behold, they spied a band of men; and they cast the man into the sepulchre of Elisha: and when the man was let down, and touched the bones of Elisha, he **revive**d, and stood up on his feet.

John 14:26 But the Comforter, which is the **Holy Ghost**, whom the Father will send in my name, he shall teach you all things, and bring all things to your remembrance, whatsoever I have said unto you.

Throughout life, you are more likely to be liked than respected. It does not require much for someone to like you. On the other hand, respect will require an honorable lifestyle of highly esteemed character . . . Love is the most costly gift, because it is given without any cause for repayment. Let God be true, and every man a liar. Do the Will of God, and thereby remain in divine focus.

The Kingdom of God is not in word, but in Power! Approaching opposition and conflict may mean to proceed in battle; rather than redirecting or fleeing for safety. The gospel of the Kingdom must be preached, and let everyone press into it!

Romans 14:17 For the **kingdom** of **God is not** meat and drink; but righteousness, and peace, and joy in the Holy Ghost.

1 Corinthians 4:20 For the **kingdom** of **God is not** in word, but in power.

Kingdom Building

Unity Effect

**Interconnected components must work and
fit interpedently as one whole unit**

The people of God should support, interact, exchange, and collaborate with other components of open silos that are aligned spiritually with each other. This works to provide one another with: faith services, fellowship, communications, network support, Christian business, financial resources, family services, mentoring, jobs, schooling, basic needs, substance abuse counseling, medical services, and legal services. Interconnected components for Kingdom Building would consist of unification of both Christian faith members and those of the Jewish faith. Believers are to unite together for the purpose of growing and operating under the fellowship and principles of the Heavenly Father. God is Lord of lords and King above all kings. His people are called to obey God's Kingdom Order, and operate within His Kingdom Dynamics that are governed by His Holy Spirit, according to His Word.

    The evangelistic objective should be preaching the Kingdom that's available through the mission of salvation to reconcile, recover, restore, develop, and enhance the capacity of each other. My eyes of faith behold the blessings of the LORD! We are made to be a blessing . . . Who have you been sent to be a blessing to? God has already ordained blessings specifically designated for you. Blessings that are so great; that even you won't have room enough to receive it all for just yourself. From out of your predetermined abundance will overflow a portion of its surplus to be shared with others . . . Believers can minister to the inner and outer man, in an effort to bring wholeness to the mind, body, and spirit to align itself in Christ.

Each component is a part of the sum of one whole unit in Christ.

Ministry + Fellowship + Services + Support + Resources + Network + Supply + Provisions = 1 Whole Unit

All components being in alliance and working harmoniously with one another as one Body.

Galatians 6:10 As we have therefore opportunity, let us do good unto all men, especially unto them who are of the **household** of **faith**.

Kingdom Building
Wrap-Around Model

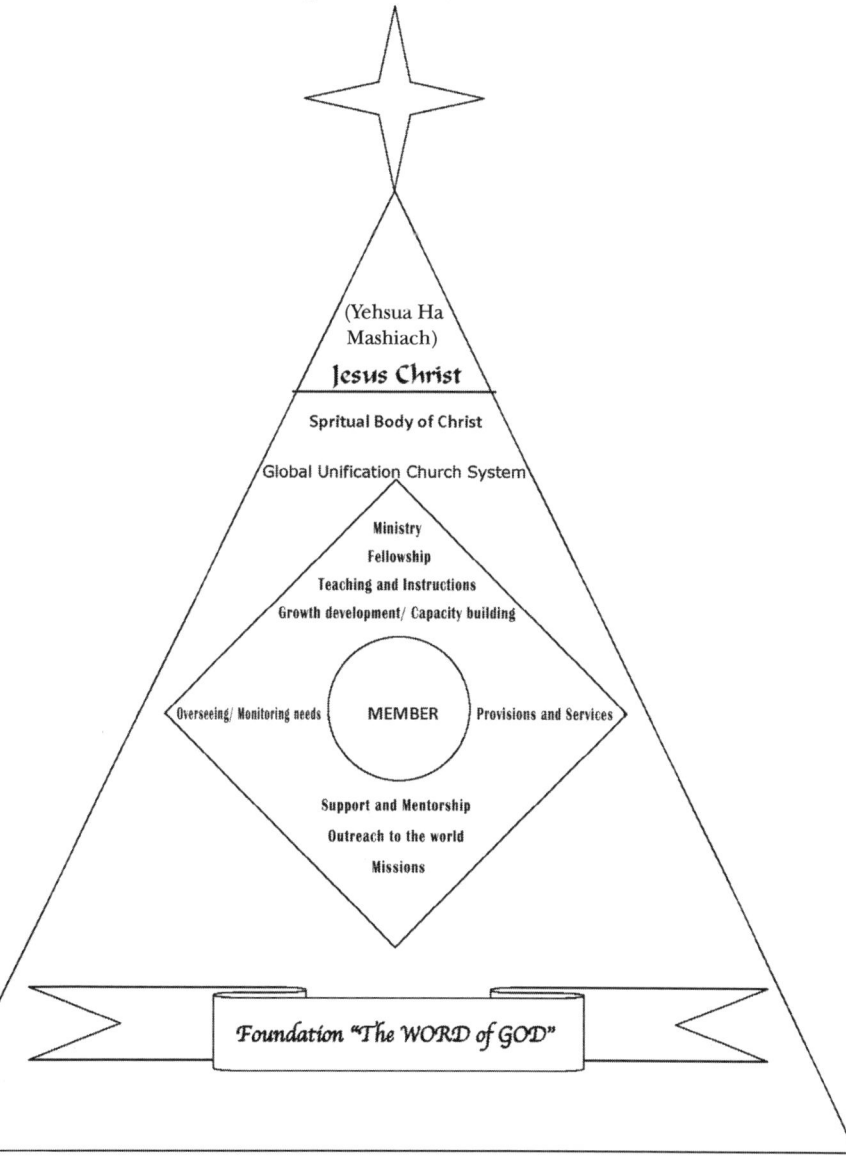

(Yehsua Ha Mashiach)
Jesus Christ

Spritual Body of Christ

Global Unification Church System

Ministry
Fellowship
Teaching and Instructions
Growth development/ Capacity building

Overseeing/ Monitoring needs | MEMBER | Provisions and Services

Support and Mentorship
Outreach to the world
Missions

Foundation "The WORD of GOD"

This support system must be empowered by the Holy Ghost, through the Word of God. Jesus Christ (Yeshua the Messiah) is the Supreme Head over the entire Global Church Body. The Church is made strong by remaining true to the integral foundation and establishment by adhering to God's principles. Therefore, we are able to pursue our appointed purposes in Christ Jesus. Each member being interdependent, according to God's divine Order and Plan. Believers will also have all goals defined and visions declared openly. A humble and submissive servant of God, will remain under righteous positioning before God and obtain grace, knowledge, wisdom and anointing for victorious living.

My prayer for the beloved people of God...

I pray that this book has presented to each reader a special message of truth. And that you would grasp a deeper and more knowledgeable understanding of how the Holy Spirit expresses Himself within us, as the Body of Christ. That you may develop a better understanding and appreciation for how each of us are uniquely designed to be who we are, and how we are set aside for a particular purpose as it pleases God. So be encouraged, those who are called to be vassals of the Most High God, and the tabernacle of His Praise!

There is power in our sincere praise to the Lord. Our God dwells within the midst of our praises to Him. In fact, the joy of the Lord is our strength. True spiritual worship draws us unto the presence of the Heavenly Father.

Index

A

Abel (son of Adam and Eve), 142-44
abominable acts, 150
Abraham (Old Testament patriarch), 37, 90, 147
acidophilus, 119
Acts
  1:5, 190
  2:1, 21
  2:36-39, 137
  2:46, 21, 26
  5:12, 21
  8:18, 180
  8:19, 180
  18:24-27, 60
  28:8, 180
Adam (first man), 80, 143, 152-53, 157, 171
adultery, 149
affliction, 115-16
agape, 162-63
ailment, 116
allergic reaction, 46
allopathic care, 126

Alpha Centauri, 129
alternative medicine, 125
Amos 3:3, 148
anger, 175-76, 187
ankle, 43, 102, 104-6
antigen, 46, 119
antihistaminic drugs, 46
aorta, 75
Apollos (Jewish man), 60
apostolic order, 17, 82, 107
appendix, 92
Aquila (Christian missionary), 60
arms, 43-44, 97-99, 185
arterial carotid, 75
artery, 86

B

bacteria, 113-14, 119, 126
balance, 55, 67, 75, 107, 119, 122, 154-55
baptism, 40, 136, 139, 170, 191, 196-97
believer, 47, 53, 62-63, 68, 70, 80, 129, 148

authority of, 199
  signs of a, 191-92, 198-99, 199-200
biceps, 98
big bang theory, 130
biochemistry, 46
bioremediation, 119
biorhythms, 128-29
bladder, 93, 95
blood, 85-86
blood offering, 142-43
blood vassal, 86
Boaz (second husband of Ruth), 147
Body
  being good to the, 120
  coming together of, 135
  importance of members, 108, 108-9
  working as one, 176
  work of, 138
bones, 81
brain, 43-44, 130

C

Cain (son of Adam and Eve), 142-44
calf, 104
Calvary's cross, 41
cancer, 85, 116, 175
celestial backside, 102
celestial bones, 81
celestial nervous system, 88
celestial nose, 69
celestial tissue, 85
celestial tongue, 71
celestial tonsils, 75
celestial waist, 101
celibacy, 151
cell, 44, 46-48, 85-88
cell membrane, 86, 88
central nervous system, 44
cervix, 95-96
Christ
  blood of, 41-42, 86, 105, 133, 135
  bride of, 15, 30, 32, 37, 65, 74, 154, 157, 191
  mind of, 33, 45, 51-52, 54, 62, 66, 80, 90, 132, 135, 169-70, 186
church, 22, 30, 32, 39, 41, 43-46, 48, 67-68, 72, 74, 85, 87-88, 135-36, 139-40, 153-57, 178-79
church cell membrane, 88
church nucleus, 87
church protoplasm, 87
cilia, 95
comparison, 43, 45
conventional knowledge and wisdom, 54
conventional medicine, 125
corporate believers, 72-73
corpus, 95
cytoplasm, 86-88
cytoplasm church, 88

D

Daniel
  2:42, 107
  7:13, 14, 82
  7:27, 31, 90
David (king of Israel), 18, 143, 165
Delilah (mistress and betrayer of Samson), 148
denominational divide, 39, 39-40, 57, 57-60
destiny, 97
Deuteronomy
  22:22-29, 149
  23:17, 149
  30:6, 90
devil. *See* Satan
diaphragm, midriff, 84
diarrhea, 126-27
digestive tract, 74, 77
disease, 114, 116, 124, 126, 175
distress, 115

divorce, 158
DNA (deoxyribonucleic acid), 65, 117-18, 134
doctrinal disagreements, 59
double helix, 118

E

eardrum, 67-68
ears, 67-68
ecclesiastical ethics, 200
elbows, 98-99
11th dimension idea, 130
El-Shaddai, 35, 72
Ephesians
  1:5, 94
  4, 170
  4:11-13, 136
  4:15, 16, 137
  5:21, 154
  5:22, 153, 157
  5:33, 165
  6:11-12, 133
  6:12, 13, 42
Eros, 147
Essence of Life, 41, 41-42
Eve (first woman), 152
Exodus
  21:5, 6, 165
  34:30, 134
external ear, 67-68
eyelids, 66-67
eyes, 43-44, 66

F

face, 43-44
face of darkness. *See* Satan
faith, 16, 19, 66-67, 136-37, 139, 146-47, 167-68, 170-72, 177, 180, 182-85
fallopian tubes, 95
Falloppio, Gabriele, 95

fat, 85
festering anger, 175
fever, 126-27, 180
fiber, 89
fiber bundle, 89
fingers, 99-100, 107
1 Chronicles 21:23, 143
1 Corinthians
  2:16, 45
  3:9-11, 132
  7:3-5, 152
  8:1, 166
  8:3, 166
  12:12, 43, 135
  12:14, 109
  12:27, 135
  13:1, 167
  13:2, 167
  13:3, 167
  13:4, 167
  13:7, 168
  13:8, 146, 168
  13:13, 146, 168
  14:1, 147
  16:14, 147
1 John
  4:8, 162
  5:8, 191
1 Samuel
  10:9, 90
  18:1, 165
  18:3, 165
  20:4, 165
1 Timothy
  4:14, 180-81
  5:22, 181
foot, 104
forearms, 98

G

Galatians 3:28, 138

generational curse, 65
Genesis
  1:2, 64, 151-52, 194
  1:27, 151-52
  1:28, 64
  2:18, 152
  2:19, 152
  2:20, 152
  2:21, 152
  2:23, 152-53
  2:23-25, 152
  2:24-25, 151
  4:1, 144, 152
  4:12, 144
  4:13, 144
  4:14, 144
  5:2, 157
  24:67, 147
germs, 111
God
  anointing of, 119
  armor of, 41-42
  judgment of, 159, 159-61
  Kingdom of, 17, 22, 24, 27-29, 59, 103, 105-6, 153, 185, 194-95
  as love, 145, 148, 162, 166
  of love and order, 63, 66
  names and references of, 34
  pleasing, 144
God effect, 15, 166
godly nature, 62, 117
good leader, 177
gospel, 7, 30, 86, 94, 124
govern, 89
grain sacrifice, 143
grief, 115, 186

H

hand, 99
harvest, 23, 25
heart, 90
Hebrew
  6:20, 12
  7:15, 12
heel, 106
hidden agendas, 107, 112
hip, 101
hip joint, 102
holism, 125
homeopathic medicines, 127
homosexuality, 149-50
human body, 123, 147, 151-52, 173, 175
  anatomy of, 12, 43, 45, 69, 74
  compared with the spiritual body, 43-109
  elements for health and healing, 120
Human consciousness, 44
human mouth. *See* mouth
human tooth, 69
husband, 15, 30, 82, 151-52, 154-58, 165

I

Immune reaction, 119
incest, 150
infectious agent, 113
inner ear, 67-68
instep, 105
intestine, 91
investment principles, 25
Isaac (son of Abraham), 147
Isaac 9:7, 18
Isaiah 45:1, 64

J

jaw, 69-70
Joel 2:28, 28
John
  7:38, 42
  11:36, 15
John (apostle), 60, 190-91

joint, 83
Jonathan (son of Saul), 165
jugular, 75

K

kidneys, 93
kingdom building, 22, 98, 107, 201, 203
Kingdom Come Mentality, 131
Kingdom dynamics, 17, 23, 63
Kingdom Order, 11, 17, 26-29, 63, 82, 130, 139
knee, 103

L

laying hands, 179-81
leadership, 87, 99, 103-4, 107, 154, 178
Leviticus
  8:24, 101
  18:6-18, 150
  20:26, 14
life stream, 41-42
ligaments, 44, 84
liver, 91
living water. See Christ, blood of
love, 72, 83, 90, 136-39, 151-58
  expression of, 161-68
  as God, 145-47
Luke
  3:16, 190
  6:35, 165
  9:1, 62
  11:12, 73
  11:2, 31, 196
lungs, 91
lust of the flesh, 56

M

marital relationship, 151, 153, 156-57, 165. *See* also love

Mark
  1:8, 197
  5:23, 180
  10:4, 158
  16:16, 17, 198
  16:17, 54
  16:18, 180
  16-19, 193
  16:20, 193
marriage, 147-49, 151-55, 157-58
mast cell, 46
mate. *See* marriage
Matthew
  3:16, 197
  5:44, 165
  7:13, 160-61
  7:13, 14, 161
  9:35, 124
  13:36-43, 29
  16:28, 134
  17:1-2, 134
  22:37, 163
  22:38, 163
  22:39, 163
  25:44, 45, 28
Melchizedeck (priest-king of Jerusalem), 140-41
microorganisms, 111-12
middle ear, 67-68
mind, 17, 19-20, 44-45, 49-52, 56-57, 62-63, 66, 79-80, 90, 114, 169-70, 172-74, 186-87
ministry of reconciliation, 28
mouth, 69-70
M theory, 128
muscles, 45, 83-84

N

Naomi (mother-in-law of Ruth), 147
natural ears, 67
natural eye, 66, 111, 113

natural healing, 125-26
naturopathy, 126-27
neck, 75
neck bone, 75
nerve endings, 45-46, 67
nervous system, 44, 46, 88
neurotransmitters, 46-47
New Jerusalem, 82
New Testament, 34, 150
Nicodemus (Pharisee), 189, 194-95, 197
nose, 68-69
nucleus, 87-88

O

Old Testament, 34, 150, 179
Order of Melchizedeck, 140-41
ordinates, 66, 68, 103, 159, 178

P

pain, 115-16
palms, 99-100
pancreas, 92
pastor, 177-78
pathogen, 111-12
Paul (apostle), 9, 151, 180
pelvis, 93
Pentecost, 21, 64, 191, 197
Peter (apostle), 134, 137, 196
Pharisees, 57, 190, 194-95
Philippians 2:2, 20
physical attractions, 147, 149, 152
physical stimulation, 149
physiology, 46
preacher, 177. *See also* pastor
pride, 20, 59, 143, 172, 201
Priscilla (Christian missionary), 60
prostitution, 149
protoplasm, 87-88, 111

Proverbs
  5:15-20, 152
  5:17-19, 152
Proxima Centauri, 129
Psalms 119:137, 14
pyrogens, 126

R

Rebecca (wife of Abraham), 147
rebellion, 13, 63, 110, 142-43, 201
receptors, 45-47, 69, 118
redemption, 24, 37, 41-42, 80, 105, 133, 176, 199
reproductive system, 94
Revelation
  17:17, 90
  19:6, 14
  21:2, 82
  21:2 10-16, 24-27, 82
  21:9, 151
  21:10-16, 82
  21:24-27, 83
ribs, 81
righteous conduct, 79, 182, 188-89
righteousness, 105, 174, 197
RNA, 117
Romans
  5:5, 145
  5:8, 146
  8:19-24, 200
  12:5-8, 38
Ruth (Moabite woman), 9, 147
Ruth 2:4-15, 147

S

salvation, 24, 53, 80, 124, 133, 146, 169, 185, 195-96
Samson (Hebrew hero), 148
Satan, 53-54, 60-62, 64, 78, 106, 111,

113, 125, 129-30, 133, 150, 156, 158, 175, 192, 198
scientific theories, 128, 128-32
2 Chronicles
   7:14, 123, 160
   30:12, 90
2 Timothy
   2:15, 178
   4:2-5, 184
Sex, 149, 149-51, 152
shoulders, 43-44, 75, 97-98
sin, 40-41, 111-13, 123-24, 143, 145, 160
sinew or tendon, 84
skeletal frame, 37, 41
snowflake, 183
sodomize, 150
Songs of Solomon 1:12-15, 152
sorrow, 115, 151, 196
soul, 50, 53-54, 76, 79-80, 90, 141, 165
South Africa, 125
speaking in tongues, 146, 191-92, 198
spinal cord, 43-44
Spiritual acidophil, 119
Spiritual Alpha-helix, 118
spiritual anatomy, 30, 30, 37, 41, 47-48, 67, 74, 78, 94
spiritual ankle, 104
spiritual aorta, 76
spiritual appendix, 93
spiritual arms, 98
spiritual arterial carotid, 76
spiritual bacteria, 113
spiritual balance, 51, 56
spiritual baptism, 197
spiritual biceps, 98
spiritual bioremediation, 119
Spiritual Bipolar, 52-53
spiritual bladder, 93
spiritual blood, 85-86
spiritual blood and water, 42

spiritual blood vessels, 86
spiritual Body, 12-13, 15, 19, 27, 30, 32-33, 37, 42, 112-13, 116
   anatomy of, 32, 37, 79, 101, 113
   compared with the human body, 43, 43-109
   elements for health and healing, 120
spiritual body arteries, 86
spiritual brain, 45, 66
spiritual calf, 104
spiritual cancer, 116
spiritual cell unit, 87
spiritual consciousness, 44, 54
spiritual demonstrations, 179
spiritual diaphragm, midriff, 84
spiritual digestive tract, 77
spiritual DNA, 65, 117-18, 134
spiritual double helix, 118
Spiritual Eardrum, 68
spiritual ears, 67-68
spiritual elbow, 99
Spiritual External Ear, 68
spiritual eyebrows, 67
spiritual eyelids, 67
spiritual eyes, 66
spiritual fat, 85
spiritual feet, 105
spiritual finger and toe nails, 107
spiritual foot, 104
spiritual forearms, 98-99
spiritual genetic coding, 80, 118
spiritual germs, 111
spiritual gifts, 76, 138, 146, 166, 171
spiritual gums, 70
spiritual hands, 99-100
spiritual headship, 45, 66, 87-88
spiritual heart, 90
spiritual heels, 106
spiritual hip, 101
spiritual hip joint, 102

Spiritual immune reaction, 119
Spiritual Infectious Agent, 113
Spiritual Inner Ear, 68
spiritual instep, 106
spiritual intestine, 91
spiritual jaw, 70-71
spiritual jugular, 76
spiritual kidneys, 93
spiritual knee, 103
Spiritual Knowledge and Wisdom, 54
spiritual legs, 102
spiritual ligament, 84
spiritual liver, 91
spiritual microorganism, 112-13
Spiritual Middle Ear, 68
spiritual mouth, 69
spiritual muscles, 83
spiritual musculoskeletal system, 44
spiritual neck, 76-77
spiritual neck bone, 77
spiritual nervous system, 47
spiritual neurotransmitters, 47-48
spiritual palms, 100
spiritual pancreas, 92
spiritual pathogen, 48, 112
spiritual pelvis, 93
spiritual perception, 45-46
spiritual receptors, 46-47
spiritual reproductive system, 94, 118
spiritual ribs, 81-82
spiritual riches and wealth, 22
spiritual RNA, 118
spiritual senses, 32, 47, 147
spiritual sinew or tendon, 84
spiritual soul, 80
spiritual spleen, 92
spiritual stomach, 78, 140
spiritual teeth, 70
spiritual thigh, 103
spiritual throat, 75
spiritual thymus gland, 76
spiritual toes, 106

spiritual trachea, 77
Spiritual Transmitters, 48, 118
spiritual triceps, 98
spiritual womb, 95-97
spiritual wrist, 99
spleen, 92
statutes, 45, 74, 159-60
stomach, 77
string theory, 128
synapse, 47

T

tares, 119
thigh, 102-3
throat, 69, 74-75
Thymus Gland, 75
tissues, 83-85, 95
Titus 2:7, 141
toe, 76, 106-7
toe nails, 107
tonsils, 75
toxicants, 110
toxins, 110-11
trachea, 69, 75
transformation factor, 117
triceps, 98
triune Godhead, 27, 33, 130

U

unbelief, 52, 63, 111, 113, 187
uncontrollable anger, 175
unity effect, 203

V

vagina, 95-96
vassal, 86
vestibule, 67-68
vindictive activities, 187

W

waist, 101
wife, 147, 151-52, 154-58, 164-65
witchcraft, 52
womb, 95-97, 194
worldly system, 61, 63
wrist, 43, 98-100

Printed in Great Britain
by Amazon.co.uk, Ltd.,
Marston Gate.